Industrial Air Pollution Monitoring

Environmental Management Series

Edited by **Professor J. Cairns, Jr**, University Center for Environmental and Hazardous Materials Studies, Virginia Polytechnic Institute, USA

and

Professor **R.M. Harrison**, Institute of Public and Environmental Health, University of Birmingham, UK.

The current expansion of both public and scientific interest in environmental issues has not been accompanied by a commensurate production of adequate books, and those which are available are widely variable in approach and depth.

The *Environmental Management Series* (formerly the *Pollution Monitoring Series*) has been established to meet the need for a set of in-depth volumes dealing with environmental issues, particularly with regard to a sustainable future. The series is published by Chapman & Hall (previously published by Elsevier Applied Science). Over a period of years, it has been building up to form a library of reference books covering most of the major topics within this diverse field.

The level of presentation is advanced, aimed primarily at a research/consultancy readership. Coverage includes all aspects of environmental science and engineering relevant to evaluation and management of the natural and human-modified environment, as well as topics dealing with the political, economic, legal and social considerations pertaining to environmental management.

Already published

1. *Instrumental Analysis Of Pollutants* (1991) C.N. Hewitt (ed.) ISBN: 1–85166–548-X.
2. *Pollution Of The Arctic Atmosphere* (1991) W.T. Sturges (ed.) ISBN: 1–85166–619–2.
3. *Organic Contaminants In The Environment* (1991) K.C. Jones (ed.) ISBN: 1–85166–621–4.
4. *Methods Of Environmental Data Analysis* (1992, reprinted 1995) C.N. Hewitt (ed.) ISBN: 0–412–73990–9.
5. *Atmospheric Acidity* (1992) M. Radojevic and R.M. Harrison (eds) ISBN: 1–85166–777–6.
6. *Global Atmospheric Chemical Change* (1993, reprinted 1995) C.N. Hewitt and W.T. Sturges (eds) ISBN: 0–412–53870–9.
7. *Biomonitoring Of Trace Aquatic Contaminants* (1992, reprinted 1994) D.J.H. Phillips and P.S. Rainbow (eds) ISBN: 0–412–53850–4.
8. *Industrial Air Pollution Monitoring* (1997) A.G. Clarke (ed.) ISBN 0–412–63880–0.

Topics currently in preparation include the following

- *Major Chemical Accidents and their Impact on Terrestrial Ecosystems.* A. Meharg. ISBN: 0–412–63410–4.
- *Artificial Reefs.* A. Jensen, K. Collins and A. Lockwood (eds). ISBN: 0–412–60370–5.
- *Biology of Pollution in Major River Systems.* T. Langford. ISBN: 0–412–80780–7 (Hb), 0–412–48170–7 (Pb).
- *Air Pollution Handbook,* 3rd edition. R.M. Harrison and C.N. Hewitt (eds). ISBN: 0-412-81640-7.

Industrial Air Pollution Monitoring

Edited by

Andrew G. Clarke

Department of Fuel and Energy,
University of Leeds,
UK

CHAPMAN & HALL

London • New York • Tokyo • Melbourne • Madras

Published by Chapman & Hall, 2–6 Boundary Row, London SE1 8HN, UK

Chapman & Hall, 2–6 Boundary Row, London SE1 8HN, UK

Chapman & Hall GmbH, Pappelallee 3, 69469 Weinheim, Germany

Chapman & Hall USA, 115 Fifth Avenue, New York, NY 10003, USA

Chapman & Hall Japan, ITP-Japan, Kyowa Building, 3F, 2-2-1 Hirakawacho, Chiyoda-ku, Tokyo 102, Japan

Chapman & Hall Australia, 102 Dodds Street, South Melbourne, Victoria 3205, Australia

Chapman & Hall India, R. Seshadri, 32 Second Main Road, CIT East, Madras 600 035, India

First edition 1998

Typeset in 10/12pt Palatino by Saxon Graphics Ltd

Printed in Great Britain by TJ International, Padstow, Cornwall

ISBN 0 412 63390 6

A catalogue record for this book is available from the British Library

Library of Congress Catalog Card Number: 97-76819

∞ Printed on permanent acid-free text paper, manufactured in accordance with ANSI/NISO Z39.48-1992 and ANSI/NISO Z39.48-1984

Contents

Contributors

Dr Andrew G. Clarke (Editor)
Senior Lecturer, Department of Fuel and Energy, University of Leeds, Leeds LS2 9JT

Mr George Bartle
(Recently retired from H.M. Inspectorate of Pollution, Leeds), 17 Hesketh Lane, Tingley, Wakefield, West Yorkshire WF3 1AU

Mr Colin Blackmore
(This work was completed when the author was Area Sales Manager, Analytical Developments Co. Ltd.), Southern UK Sales Manager, Monitor Labs Ltd., 2 Airfield Way, Christchurch, Dorset

Dr Bernard M. Gibbs
Senior Lecturer, Department of Fuel and Energy, University of Leeds, Leeds LS2 9JT

Dr Nigel Gibson
Senior Consultant, Room 3.12, E1 Building, AEA Technology, Culham, Abingdon, Oxfordshire OX14 3DB

Dr Richard Grant
Application Engineer, Combustion Developments Ltd (CODEL), Station Building, Station Road, Bakewell, Derbyshire DE4 1GE

Mr Richard Horne
Sales Manager, Combustion Developments Ltd (CODEL), Station Building, Station Road, Bakewell, Derbyshire DE4 1GE

Mr Simon Medhurst
Environmental Quality Manager, CRE Group Ltd., Stoke Orchard, Cheltenham, Gloucestershire GL52 4RZ *and* Quality Officer, Source Testing Association

Mr David P Miles
Director, Environmental Evaluation Ltd., Lawton Square, Delph, Oldham OL3 5DT

Dr Stephen Richardson
Senior Consultant, AEA Technology, National Environmental Technology Centre, Culham, Abingdon, Oxfordshire OX14 3DB

Mr John Scawin
(Recently retired from British Oxygen (Special Gases) Ltd), 51 Merland Rise, Tattenham Corner, Epsom KT18 5RN

Mr John Turnbull
Sales and Applications Manger for Stationary Emisions Monitoring, Signal Instruments Company Ltd., Standards House, Doman Road, Camberley, Surrey GU15 3DW

Mr David S Walker
(Recently retired from Regional Manager, Glasgow Office, AEA Technology), 98 Aitken Road, Hamilton, Lanarkshire ML3 7YL

Mr Ian Watson
Regional Sales Manager, Northern Britain and Europe, Land Combustion Ltd., Dronfield, Sheffield, South Yorkshire S18 6DJ

Dr Paul T Williams
Senior Lecturer, Department of Fuel and Energy, University of Leeds, Leeds LS2 9JT

Series editors' foreword

Focus is now shifting from concern over the toxicological aspects of pollutant discharges and waste disposal to the larger issue espousing a transition to new resource use policies that: conserve natural resources and energy for long-term sustainable use; minimize ecological damage during the extraction of raw materials; minimize wastes during production and recycle as much as possible of the wastes produced; facilitate the re-incorporation of the product into natural systems at the end of its life cycle; use wastes from one production process as inputs (i.e. raw materials) in some other production process (e.g. municipal sludge to agricultural production). There is a concomitant shift from merely preventing observable deleterious effects from potentially toxic materials to enhancing ecosystem health and condition, which will probably involve a major shift from a laboratory focus in toxicology to an ecosystem focus because of the difficulty of simulating some of the critical attributes in microcosms and mesocosms. The problem of interfacing two complex multivariate systems – one technological, the other ecological – will be best resolved by developing a process characteristic of the information age of integrative thinking and reasoning.

Environmental problems transcend the capabilities of a single discipline, although many disciplines will have much to contribute. The series is a forum on the interface between technological systems and ecological systems with the recognition that both are part of human society's life support system. Some of the books in the series will be on methodology, particularly those methods suitable for generating information that will affect the decision making process. We also feel that policy and management, in addition to science and technology, have a role in the overall mix. All of the volumes in the series will have a single unifying theme – the intelligent use, without abuse, of environmental resources. Because the series is a forum, opposing views are welcome and desirable if kept within normal bounds of professional ethics.

The Series will approach these objectives by publishing both single-

and multi-author books and edited volumes in all aspects of environmental science and engineering pertinent to understanding and managing the physical, chemical and biological environment as well as topics dealing with the political, economic, legal, and social considerations. The level of presentation will be advanced, the books being aimed primarily at a research/consultancy readership.

J. Cairns, Jr
R. M. Harrison

Preface

Since the introduction of the Environmental Protection Act 1990 in the UK there has been rapid growth in the amount of monitoring of air pollutant emissions from industrial processes. Many industries which in the past did not have to measure the emissions from their chimneys now find themselves either installing unfamiliar and expensive equipment or paying contractors to make measurements on their behalf. Whilst in other countries, particularly the USA and Germany, standard methods of measurement had been established for some years, the UK found itself trying to implement the new legislation with very few standard methods or protocols. Although the situation is improving, there is still a shortage of information on the various methods of pollutant monitoring and their relative advantages and disadvantages. It is hoped that this volume will help all those involved – industry, the control authorities, consultants, instrument suppliers etc., to gain a broad overview of the topic. Readers wanting detailed instructions for particular methods will be disappointed – there is simply not room in a book of this size to go into that level of detail. Hopefully, sufficient references have been given for those wanting further details to know where to find them.

Emissions monitoring is far more than just choosing the right equipment or methods. There are many issues such as the nature of the emissions limits, the legal requirements, the collecting and processing of the data, quality assurance and control, calibration and accuracy, safety and the management of site work. We have attempted to cover all of these subjects. So, although I am based in a University, this is intended to be a practical rather than an academic book, firmly grounded in the real world.

The immediate impetus for producing the book was the 3-day short course on Industrial Air Pollution Monitoring run at the University of Leeds since 1991. I am grateful for several of the lecturers agreeing to upgrade their notes into a chapter or part of a chapter and to those who have not been involved in the course but agreed to make contributions.

The origin explains the structure of the book – a relatively large number of fairly short chapters. It also explains any differences in the style and level of detail between different chapters.

Details of companies who manufacture or distribute equipment for emissions monitoring are not incorporated here, but such information is available in a Directory of Emissions Monitoring produced biennnially in conjunction with our short course. Copies can be obtained from Miss J. Charlton, Short Course Manager, in the Editor's Department.

Finally, my thanks to the members of our secretarial staff, Heather, Sheelagh and Julie who in in various ways have contributed towards the preparation of the manuscript.

Andrew Clarke

1

Legislation on emissions of air pollutants and their monitoring

Andrew G. Clarke

1.1 WHY MONITOR EMISSIONS?

This book is concerned with the monitoring of the emissions of air pollutants into the atmosphere from industrial processes. The reason why this topic has become such an important one in many countries is that there is an increasing weight of legislation requiring operators to provide information about gaseous and particulate substances which are released into the environment from the processes they control. The concerns which lie behind the legislation may relate to

- human health effects of pollutants;
- potential damage to other aspects of the environment (for example acid rain and ozone generation with their consequent effects, or building decay);
- amenity issues (for example black smoke or visibility degradation); or
- global issues such as climate change.

The legislation often takes the form of emissions limits or guidelines below which it is expected that the process will operate. Demonstrating compliance with these limits is then the reason for the monitoring effort. In the next section we shall examine the approach to emissions limits in several countries.

As will become apparent through the reading of this book there are many problems associated with the obtaining of reliable monitoring data. It is therefore preferable to use methods and equipment which have been tested and approved. The third section of the chapter surveys the extent to which appropriate guidance in these areas is available.

The measurement of pollutant concentrations in waste gas streams is undertaken for many good reasons apart from the obvious one of it being a legal requirement. Here are a few of them:

1. **Assessment of process efficiency**. Emissions may be of the target product of a process which has not been captured, or of substances which are potentially convertible to product. For example, NO_2 emissions from nitric acid manufacture, SO_2 emissions from sulphuric acid manufacture, or volatile organic compounds from various chemical processes. Losses up the chimney in each case represent a reduction of product yield or reduction in overall process efficiency.
2. **Process control**. Feedback from a continuous monitor may be employed in a process control system. An example would be the use of smoke or carbon monoxide monitors for control of the air supply in a batch incinerator.
3. **Determination of efficiency of a pollution control device**. Pollution control devices such as scrubbers or electrostatic precipitators will often be specified in terms of the percentage removal of the input material rather than in terms of absolute emissions. Commissioning and acceptance testing will involve the measurement of pollutant concentration before and after the device. Similar tests will also be necessary if the device appears to be inadequate for the task or to be underperforming later in its life. Routine monitoring of the performance of abatement plant requires continuous indication of emissions levels. For example, bag failure in a filtration unit needs to be detected immediately.
4. **Collecting data for emissions inventory compilation**. Emissions inventories, whether on a national regional or local basis now form an essential component of air quality management. Data which in the past have been estimated from uncertain emissions factors applied to a process will increasingly be gained from actual measurement. For example, national NO_x and SO_2 data are needed in relation to the acid rain problem. While SO_2 can be estimated with reasonable accuracy from regular fuel sulphur analysis, this is not the case with NO_x, which varies considerably with both fuel nitrogen content and combustion conditions. Continuous monitors will provide much improved information. Another example would be the new chemical release inventory which is now being built up in the UK by the Environment Agency [formerly H.M. Inspectorate of Pollution (HMIP, 1996)] or the Toxics Release Inventory in the USA.
5. **Data for formal environmental impact assessment (E.I.A.)**. E.I.A. is now generally required as part of the authorization or permitting process for new plant. The emissions data must in this case be based on previous experience with similar plant. Continued authorization

of existing plant may also require an environmental impact assessment and this must be based on actual emissions data from the plant.

6. **Data for internal company environmental audits**. Assessment of performance by comparison with legal standards, guideline values or environmental good practice elsewhere requires detailed knowledge of current emissions.
7. **Check monitoring on behalf of the regulators**. Auditing of returns of monitoring data to the Environment Agency will be accompanied by a programme of check monitoring to ensure that the data are accurate.
8. **Concern for in-plant safety of personnel and materials**. Some pollutants have a direct impact on the components or materials of construction of the plant – for example corrosive acid gases such as HCl or H_2SO_4 vapour. Design must take account of these impacts and operating temperatures may need to be adjusted if acid dewpoint problems are encountered. Other pollutants may present hazards to the workforce in the event of plant failure or leakage. A small boiler or furnace fired with waste oil containing significant levels of lead might be a typical case.
9. **Evaluation of non-regulated emissions of potential concern**. The knowledge or perception of the hazard associated with particular pollutants tends to evolve with time and new concerns regularly come to the fore. Over the past few years the emissions of polychlorinated dioxins and furans have been the focus of particular interest, although the setting of rigid limits in various countries has progressed rather slowly. Partly this has been due to the medical uncertainties associated with the effects of minute levels of these compounds, partly due to the severe analytical problems, and partly due to uncertainties over the formation mechanisms and possible control strategies.
10. **Investigation of problems**. A typical example would be monitoring arising from local complaints about odours in the vicinity of a chemical works.

1.2 EMISSIONS LIMITS

It is important to understand the legal framework within which emissions limits are set and implemented. This framework obviously varies from country to country and reflects the structure of the political institutions and the balance of power between national and regional or state authorities. The stringency of legislation will reflect the strength of the commitment to environmental protection. The following sections give a brief summary of the legal and administrative positions in the European

Union, the UK, Germany and the USA. The actual nature and implications of emissions limits are examined in section 1.2.6

1.2.1 European Union legislation

Following the Maastricht Treaty, the European Community (EC) formally became the European Union (EU) and the Council of Ministers was retitled the Council of the European Union. Measures to control pollution are drawn up by the European Commission and, after discussion, are approved by the Council. They are usually issued in the form of *Directives* which set out the objectives and the standards to be achieved. Each member state must then draw up regulations within their own legal framework to comply with the terms of the directive.

For combating air pollution from industrial plants a general or 'framework directive' was passed in 1984 (directive number 84/360/EEC). This requires the process of authorization for various categories of industry and requires the use of BATNEEC (Best Available Technology Not Entailing Excessive Cost). Note that in the UK the word Techniques is substituted for Technology. The directive has been implemented in the UK by the Environmental Protection Act 1990 which is discussed below.

Specific directives for particular types of plant have followed the framework directive and these include three dealing with incinerators (new Municipal Solid Waste plant 89/369/EEC and existing MSW plant 89/429/EEC; hazardous waste plant 94/67/EC) and one with large combustion plants (88/609/EEC). The latter controls emissions of SO_2, NO_x and ash from plants with energy input of over 50 MW (primarily but not exclusively electricity generation). Emissions limits for individual new plants are given, but in addition there is a requirement that each member state must reduce its national emissions of SO_2 and NO_x by certain amounts relative to 1980 over a set time scale. For the UK, the former West Germany (FRG) and France these reductions in percentage terms are

Year	1993			1998			2003		
	UK	FRG	France	UK	FRG	France	UK	FRG	France
SO_2	20	40	40	40	60	60	60	70	70
NO_x	15	20	20	30	40	40			

The Directive requires that monitoring of the relevant pollutants is undertaken at the operator's expense, that the monitoring techniques represent best industrial measurement technology and that they are approved by the regulating authority. Calibration and maintenance are also covered. The directive also requires emissions inventories to be set up so that total annual emissions from each plant must be obtained.

Further details of the incineration directives are given below. Additional directives on smaller combustion plant (10 to 50 MW) and gas turbines are anticipated in the near future and a revision of the Large Combustion Plants Directive affecting the 50 to 100 MW size is planned. Another directive (96/61/EC) introduces Integrated Pollution Prevention and Control (IPPC) which will modify the current UK Integrated Pollution Control (IPC) regulations discussed below.

1.2.2 UNECE protocols

The United Nations Economic Commission for Europe (UNECE) comprises all the countries of Europe and North America. In 1979, the Commission adopted the Convention on Long-Range Transboundary Air Pollution which had as its main focus at that time the problem of acid rain. The Convention says that the countries should 'endeavour to limit and, as far as possible, gradually reduce and prevent air pollution, including long-range transboundary air pollution'. Requirements relating to specific pollutants are set out in protocols to the Convention. Three specific protocols cover sulphur emissions, nitrogen oxide emissions and volatile organic compounds.

Neither the UK nor the European Community ratified the sulphur protocol (1985) which required signatories to reduce national sulphur emissions by 30% by 1993 relative to 1980 levels. The Large Combustion Plants Directive discussed above was the corresponding EC action. However, a revised protocol was negotiated and signed in 1994 and both the UK and the European Union have signed this. Individual countries have committed themselves to national reductions of SO_2 emissions relative to 1980. For the UK the figures are 50% by 2000, 70% by 2005 and 80% by 2010, which is rather slower progress than in some other European countries.

The 1988 Nitrogen Oxides Protocol freezes NO_x emissions by 1994 using a 1987 baseline. This was ratified by the UK and by the EC. The approach to both sulphur and nitrogen pollutants is now based on the critical loads approach and critical loads maps are being drawn up for the whole of Europe based on data produced by the individual countries for the European Monitoring and Evaluation Programme (EMEP, which was launched by a 1984 LRTAP Protocol). Future national emissions reductions will be negotiated in the light of these critical loads assessments.

The Volatile Organic Compound (VOC) Protocol was signed in 1991 obliges most parties to achieve 30% reduction of emissions of non-methane organic compounds by 1999 using 1988 as the base year, although alternative base years can be adopted. The protocol is wide-ranging covering the setting of emissions standards for VOCs, promotion of products with low or zero solvent content, the adoption of best available technologies for emissions reduction, especially where ozone levels are high, and adoption of methods to reduce emissions from gasoline handling and storage. As part of the response to this protocol in 1992 the EC agreed a first stage directive relating to gasoline storage and distribution and a solvents directive is expected to be agreed shortly. In the UK, any emissions limits or other requirements which arise from these actions will be effected through the Environmental Protection Act or other secondary legislation.

1.2.3 UK legislation

Background

Historically, air pollution control in the UK has been divided between a national inspectorate responsible for the larger and most polluting sources and the Local Authorities (Environmental Health Departments of City or Borough Councils) who were responsible for small to medium sources of pollution, mostly combustion processes. The powers available to Local Authorities were very limited and of a different nature to those available to the inspectorate. They included various provisions to control the smoke nuisance including the establishment of 'smoke control zones' where the burning of bituminous coal in homes was restricted. The control of gaseous emissions was restricted to controls on the siting of new plant through planning regulations, the amelioration of their effects via chimney height determination, and controls of public nuisances under the Public Health Act. Emissions limits supervised by the Local Authorities were restricted to the area of smoke, grit and dust control.

The national inspectorate controlled emissions from all the larger and potentially more polluting sources. The basic requirement was for the operators to use Best Practicable Means (BPM) to reduce the emissions or, failing that, to render them harmless. Presumptive limits on emission were then set – i.e. limits which, if adhered to, was evidence supporting the presumption that BPM was being used.

The Environmental Protection Act 1990 radically changed the whole system of control. Some of its requirements were designed to bring UK legislation into line with European Union Directives discussed above. The Act maintained a division of responsibilities for air pollution control between Her Majesty's Inspectorate of Pollution (HMIP) and Local Authorities and these are summarized below.

On 1st April 1996, HMIP was merged into a larger Environment Agency which also includes the National Rivers Authority and the Waste Regulatory Authorities. The basis of the new organization was laid down in the Environment Act 1995. *References in this book to HM Inspectorate of Pollution should be understood now to refer to the Environment Agency.*

The situation in Scotland and Northern Ireland is distinct from that in England and Wales and each country has its own control body. In Scotland there is a new Scottish Environmental Protection Agency which holds the authority to regulate all the air emissions which in England and Wales would be split between the Local Authorities and the Environment Agency.

The residue of legislation from the Clean Air Acts 1956 and 1968 and the Control of Pollution Act 1974 not already repealed in the Environmental Protection Act 1990 (which includes the older emission limits for plant below 20 MW thermal input) have been consolidated with minor changes in the Clean Air Act 1993.

National control

Integrated Pollution Control is applied to a group of *prescribed processes* (the list of processes is described as Schedule A). The prescribed processes include those that involve major air pollutant emissions, those that involve aqueous discharges of particularly toxic substances, or those giving rise to significant quantities of special waste (i.e. solid wastes of a more toxic or hazardous nature). Integrated Pollution Control means that the control of the potential pollution caused by gaseous, liquid and solid wastes is dealt with in an integrated way. The aim is to develop an approach to pollution control that considers discharges from industrial processes to all media (air, water, land) in the context of the effect on the environment as a whole. This avoids the danger of solving an air pollution problem (for example) in a way which results in large amounts of liquid effluent or solid waste.

Operators must use the BATNEEC to prevent or minimize the release of prescribed substances into the environment and to render harmless any such substances which are released. The first requirement is to prevent the release and if that is not practicable then the release should be rendered harmless – for example, by dispersion through a tall chimney. The prescribed substances for release into the air are given in Box 1.1.

The term BATNEEC derives from European legislation as explained above and replaces the phrase 'best practicable means' which was previously used in the UK. The Best Available Techniques for control will normally be specified by the Environment Agency (formerly HMIP) following discussions with the industries concerned. Detailed guidelines are given in Process Guidance Notes (one set of notes for each process) which are issued and updated every 4 years. These notes may cover such topics as

- Emissions limits for various pollutants from new and existing plant
- Monitoring and reporting requirements
- Process operating conditions
- Pollutant removal technology
- Chimney requirements (e.g. efflux velocity)

Box 1.1 Prescribed substances for release into the air

(Environmental Protection (Prescribed Substances and Processes) Regulations, 1991)

Oxides of sulphur and other sulphur compounds
Oxides of nitrogen and other nitrogen compounds
Oxides of carbon
Organic compounds and partial oxidation products
Metals, metalloids and their compounds
Asbestos, glass fibres and mineral fibres
Halogens and their compounds
Phosphorus and its compounds
Particulate matter

No prescribed process may be operated without an Authorization from the Environment Agency after a date specified for that type of process. The implementing of the system of authorizations has been spread over several years. The 'Fuel and Power Industries' received authorization in 1992 but the final group of processes (the 'Other Industry' group mentioned above) did not receive authorization until 1996.

The process operator must apply for authorization before a set date. The application must include, among other things, estimates of releases of pollutants to air, to surface waters, to sewers and to land, and an evaluation of environmental impacts of these releases. All the parameters that will be monitored (both relating to the process and relating to the emissions) must be described, together with details of analytical procedures, quality assurance and record keeping.

The authorization indicates conditions which must be complied with (e.g. emissions limits, fuel quality, monitoring requirements). It may give temporary authorization for continuing operation in the plant's current state but demand improvements which must be achieved within a set time scale. If an authorization is refused the operator has the right to Appeal to the Secretary of State for the Environment.

The air pollutants which must be measured and the frequency of monitoring are specified in the terms of the authorization. It is the responsibility of the process operator to provide the monitoring data to demonstrate compliance with the authorization. The operators must therefore either install appropriate continuous monitoring equipment or undertake periodic measurements of emissions themselves. Alternatively, they may contract private companies to make the measurements. Neither the Environment Agency nor the Local Authorities undertake routine monitoring of air emissions. They will ensure that the analytical procedures, quality control and record keeping are satisfactory. They have the authority to enter premises both to inspect and to take measurements, or to instruct environmental consultancy companies to do this on their own behalf. A regular programme of such 'check monitoring' is undertaken.

The new system is designed to provide much more information to the public. Copies of the application and the authorization are available to any member of the public to inspect. Details of monitoring data obtained to demonstrate compliance is also made available on a public register. Process operators can request that information be withheld from the public if it is considered to be commercially confidential. For example, this situation arose in a dispute between HMIP and the electricity generation companies who did not wish to reveal their future fuel supply plans.

Local Authority control

Local Authorities are responsible for the control of air pollutant emissions from a further group of prescribed processes (Schedule B). They do *not* apply Integrated Pollution Control and so under the Environmental Protection Act are not concerned with water pollution. The range of Schedule B processes is very wide. It covers the smaller-scale combustion, metallurgical and mineral processes (e.g. incinerators, foundries, glass works) but also covers processes which give rise to the release of volatile organic compounds (paint spraying and various coating processes). It also covers odours which arise from the processing of animal remains.

The system of air pollution control operated by the Local Authorities is now very similar to that operated by the national Agency, which was not the case before the 1990 Act. BATNEEC must be used to prevent the emissions and the expected standards are described in a further set of Guidance Notes issued by the Secretary of State. Authorizations must be sought for existing plant and for any new plant before it can start operation. Emissions limits and monitoring information are available to the public.

In addition to their responsibilities under the Environmental Protection Act the Local Authorities continue to have responsibility for the smaller combustion processes (as described previously) and they have gained new responsibilities in the 1995 Environment Act for Local Air Quality Management.

1.2.4 German legislation

Germany is a federation of states (Lander), each of which has its own parliament. The 11 'old states' of the FRG were joined by the eastern states of the former GDR after the unification treaty in 1990. Much of the implementation and enforcement of federal legislation is undertaken by the states (Boehmer-Christiansen and Skea, 1991; Murley, 1995). Compared with the UK, federal institutions are relatively weak in terms of their management powers. Responsibility for environmental affairs lies with the Federal Ministry for Environment, Nature Protection and Reactor Safety (Bundesministerium fur Umwelt, Naturschutz und Reaktorsicherheit or BMU), which was formed in 1986. Technical expertise and advice to the ministry is provided by the Federal Environmental Agency (Umweltbundesamt or UBA) which also coordinates the implementation of environmental policy between the states.

In 1974 the Federal Immision Control Act (Bundesimmisionsschutzgesetz or BImSchG) was passed. Immission is a word that has no direct English equivalent but relates to the pollution at the point of impact rather than the point of emission. This act enabled the Federal Government to pass ordinances or directives relating to air pollution, either of which may specify air quality standards or emissions limits. BImSchG specifies that emissions from plant requiring authorization should be reduced using Stand der Technik (= Best Available Technology or, loosely, state of the art). As in the UK with Best Practicable Means, the terminology has resulted in considerable debate as to its practical implications, particularly over the economic aspects.

There are basically two types of legislative control. An Ordinance (Verordnung) is directly binding on all persons and bodies within the country. An Administrative Regulation or Directive (Verwaltungsvorschrift) does not have the status of an act but directly obliges the supervisory and licensing authorities to ensure uniform implementation of statutory regulations in all the states. The terminology is similar to that used within the EU where Regulations apply directly to the relevant parties while Directives are applied indirectly within the member states. In fact, the relationship between member states and the European Union has some similarities to the relationship between the Lander and the Federal government in Germany.

Over 20 Ordinances have been issued of which one of the most significant related to Large Combustion Plant (Grossfeurungsanlagenverordnung – GFAVo) in 1983 (the 13th BImSchV – Bundesimmisionsschutz-verordnung). It sets emissions limits for SO_2, NO_x, dust, CO, halogens and carcinogens and applies to all plant over 50 MW thermal. This legislation formed the model for the draft Directive on Large Combustion Plant issued by the Commission of the European Communities in 1983. However, it was not well received, especially in the UK, and there followed protracted negotiations before the EC Directive in its final, less stringent form was agreed in 1988.

The majority of air pollution control measures, including emissions limits, derive from the administrative directive generally known as TA(Luft) (Technische Anleitung zur Reinhaltung der Luft – Technical instructions for air quality control). This has been updated several times, most recently in 1986. TA(Luft) specifies that emissions measurements should be carried out using guidelines issued by the VDI (see below). As in the USA with Federal/State relationships TA(Luft) does in principle allow Lander to set more stringent limits than the Federal ones. In practice, this is unpopular and has resulted in legal challenges relating to fair competition.

Technical guidelines on pollution control and monitoring have been prepared by the Association of German Engineers (Verein Deutscher Ingenieure – VDI). These cover both emission measurement and immision measurement (i.e. ambient air). A list of some of those relating to emission monitoring is given as an appendix. A manual covering methods for continuous emission monitoring systems has been produced by the Federal Ministry for the Environment, Nature Conservation and Nuclear Safety (1988).

1.2.5 US legislation (Code of Federal Regulations)

The modern era of air pollution control in the USA began with the Clean Air Act of 1970 and the establishment of the Environmental Protection Agency. EPA set National Ambient Air Quality Standards (NAAQS) and each state was required to develop State Implementation Plans in order to achieve compliance with the Federal goals. As part of the control strategy the Act required the EPA to set New Source Performance Standards (NSPS) for various categories of industrial plant emitting 'criteria' pollutants, i.e. those with air quality standards. States may, if they wish, bring in emissions standards which are more stringent than the national standards. These NSPSs have been promulgated and regularly revised since they were first introduced in 1971. EPA was also required to set National Emission Standards for Hazardous Air Pollutants (NESHAPS). Before the 1990 Clean Air Act Amendments the species covered were asbestos, beryllium, coke oven emissions, inorganic arsenic, mercury, vinyl chloride, benzene and radionuclides.

Part of the 1977 Clean Act Amendments dealt with the issue of the Prevention of Significant Deterioration of areas which currently comply with air quality standards and there were other new concepts which affected the way in which the emissions standards were applied. The 'bubble' strategy addressed the total emissions from a group of emitters affecting air quality in a particular area. In relatively clean areas a company may balance emissions increases at one plant with reductions at another, provided that the total emissions are not increased. The building of a new plant in a non-attainment area is only permitted if the new source emissions are 'offset' by reductions in other plant. A company shutting down or modifying a facility can 'deposit' the reduction in emissions for later use as an offset ('banking' policy) or even sell it to another company. As yet, there is no corresponding system of control in Europe; however, the Large Combustion Plants Directive effectively applies national 'bubbles' to allow flexibility in achieving emissions reductions of SO_2 and NO_x.

These approaches to air quality management continued to be applied in the 1990 Clean Air Act Amendments. The non-attainment of the ozone, carbon monoxide and particulate matter air quality standards was addressed (Title I). Depending on the severity of the problem different deadlines for compliance, major source definitions and emissions reductions will be applied. For ozone, this means targeted reductions of VOC and NO_x emissions. In relation to acid rain, major reductions of SO_2 and NO_x emissions are required (Title IV). The affected sources includes most electrical utilities that burn coal or oil and some other industrial facilities. Phase I sources are limited to 2.5 lb/MMBtu of SO_2 by 1995 while a very large number of Phase II sources will be limited to 1.2 lb/MMBtu by 2000. Emissions will be allocated to plant and made transferable on the basis of a system of allowances which may be banked or sold. One allowance equals one ton per year of SO_2 emissions. Tighter NO_x standards for utility boilers will be progressively introduced. In both phases the affected plant will be required to install CEMs. The list of Hazardous Air Pollutants is extended to 189 species (Title III). EPA has published a source category list and regulation schedule and Maximum Achievable Control Technology must be introduced. The initial MACT standards covered 41 source categories and the remainder will be introduced up to 2000. A system of 5-year operating permits for all major sources has been introduced (Title V). Each state must draw up regulations for their permit programs based on EPA guidelines and these must be formally approved by EPA. Permits must identify all sources of emissions, the required emissions standards, the schedule for compliance and monitoring requirements. The permitting system bears many similarities to the UK authorization system which in turn is based on the EU directive.

1.2.6 The nature and implications of emissions limits

1. An emissions limit applies to a specified *process*. Different processes may have different limits for the same pollutant.
2. The limit may be variable with the *size of plant* either on a continuous sliding scale (e.g. SO_2 emissions covered by the EC Large Combustion Plants Directive) or on the basis of different bands above and below specified sizes.
3. The limit will generally refer to exhaust gas concentrations corrected to *reference conditions* of temperature, pressure, oxygen concentration (sometimes CO_2 in place of oxygen) and water vapour content conditions is discussed in Chapter 2.

Table 1.1 Comparison of emission limits for municipal waste incineration (units: mg/m^3)

Country	EU	EU	EU	UK	UK	Germany	Germany	USA
Regulation		89/369/EEC 1989		IPR5/3 1992	PG5/4 1991	TA (Luft) 1986	17.BImSchV 1990	56FR5507 1991
Plant size	3 tonne/h	1 – 3 tonne/h	<1 tonne/h	>1 tonne/h	<1 tonne/h			>225 tonne/day
Ref. conditions		273K, dry gas 11% O_2 or 9% CO_2		273K, dry gas 11% O_2	273K, dry gas 11% O_2	273K, dry gas 11% O_2	273K, dry gas 11% O_2	293K, dry gas 7% O_2; or CO_2 used
Particulates	30	100	200	30	200	30	10	34 or 10% opacity
CO	100	100	100	100*	100	100	50	50 or 100 ppm†
SO_2	300	300	–	300	300	100	50	30 ppm or 80% reduction
HCR	50	100	250	30	250	50	10	25 ppm or 95% reduction
HF	2	4	–	2	–	2	1	
NO_x as NO_2	–	–	–	350	–	500	200	180 ppm
Volatile organics as total carbon	20	20	20	20	20	20	10	
Pb+Cr+Cu+Mn	5	5	–	1 (incl Sn) (Pb+Cr+Cu+Mn and Ni+As together)	5 (Pb+Cr+Cu+Mn, Ni+As and Cd+Hg together)	5	0.5‡ (Pb+Cr+Cu+Mn and Ni+As together)	
Ni+As	1	1	–			1		
Cd+Hg	0.2	0.2	–	0.1 each		0.2 (Incl Tl)	0.05 Cd+Tl 0.05 Hg	
Dioxins (TEQ) ng/m^3	–	–	–	1 Guide 0.1			Guide 0.1	30

* IPR5/3 requires that CO should not exceed 100 mg/m^3 hourly average and that 95% of all 10-min averages taken in any 24-hour period should not exceed 150 mg/m^3.
† Limit depends on design of incinerator. Those involving RDF firing have a higher limit of 150 mg/m^3.
‡ Total of Sb, As, Cr, Co, Pb, Mn, Ni, Sn, V.

4. The limits must be taken in conjunction with the *monitoring frequency* and *length of measurement*. Some species must be monitored continuously and the limit may relate to 15-minute, half-hourly or hourly averages. Other species may only need to be measured a few times a year using a set minimum sampling period.
5. The *statistical criterion* that must be satisfied is specified. For continuous monitoring this might be that 95% of the hourly readings should not exceed the limit value. Even this apparently clear criterion is in fact ambiguous. For example, do you apply it a week at a time or on a rolling basis? For measurements only required once or twice a year the criterion to be satisfied may simply be that none of the measured values should exceed the emissions limit.
6. For some pollutants emissions, *guidelines* may be set rather than limits. These represent the levels that should be achievable with best available practice, but are not legally enforced.

To illustrate these points the example of municipal waste incinerators is taken since regulations relating to incinerators generally require a large number of components to be measured. Table 1.1 indicates some EU, UK, German and US requirements. The controls cover particulate material, partial combustion products (CO and VOCs), acid gases (SO_2, HCl, HF, NO_x), metals and dioxins. The UK requirements for Part A and Part B processes (IPR5/3 and PG5/4) implement the EU directive but in some cases are more stringent, such as the addition of a NO_x limit for large sources. The list of metals to be analysed varies between the countries; for example, Germany includes thallium in the group of potentially volatile metals along with mercury and cadmium and adds vanadium, cobalt and tin to the other metals in the EU Directive. The emissions of organics are expressed in terms of total carbon which implies that non-speciated emissions measurements are undertaken using a flame ionization detector calibrated with a reference hydrocarbon. The total polychlorinated dibenzo-dioxins and dibenzo-furans (PCDDs and PCDFs) are expressed as Toxic Equivalent values (TEQ). This is explained in Chapter 10. HM Inspectorate of Pollution set a limit value of 1 ng TEQ/m^3 with a guide value of 0.1. Not included in the table are the required frequencies of measurement which vary from continuous monitoring (e.g. CO and particles by opacity) to annual measurements (e.g. dioxins).

The wording of the emissions limit document is critical to the interpretation. The following example from HM Inspectorate of Pollution in IPR5/3 illustrates the precise nature of the definitions that need to be given: 'The averaging period for all continuous monitors should be calculated by taking into account only the hours during which the plant is actually in operation, including start-up and shut-down. For continuous monitoring the release limit figures are complied with if 95% of the hourly average readings for each rolling 24 hours do not exceed the emission value and the peak hourly average value does not exceed 1.5 × the limit value'.

1.3 MONITORING REQUIREMENTS

1.3.1 Sampling protocols

The overall process of sampling and analysis requires careful attention to detail and the development of sampling protocols. Box 1.2 shows a summary given in one of the UK Guidance Notes. Many of the issues raised will be subject to detailed discussion in later chapters. The issue of representative sampling is crucial. The decisions to be taken concern

- the geographical position within the plant;
- the point (or points) within the duct or chimney where the sampling probe is to be situated;
- the time at which the sample is taken relative to process operational conditions (particularly important for batch or cyclic processes);
- the length of the sample period; and
- ensuring integrity of the sample during extraction, handling and transport.

The question of representative sampling is covered in Chapter 3 in relation to particles and Chapter 5 in relation to gases. Other issues which affect the quality of the data are discussed in Chapter 15, which covers many aspects of the management of an emissions monitoring programme.

In the rest of this chapter we focus on the question of standard methods.

1.3.2 UK standard methods

Unlike the situation in the USA or Germany, the implementation of emission limits arising out of the Environmental Protection Act 1990 has been carried forward with relatively few standard test methods. Those that exist are published as British Standards, but development of new standards is slow and it will be some years before even the most commonly measured pollutants are covered. A full list of the relevant British Standards is given in Appendix A.

HM Inspectorate of Pollution have produced several helpful Technical Guidance Notes which act as pointers to acceptable methods (HMIP, 1993a,b, 1995a,b). The note 'M2' covers available monitoring techniques and indicates many of the standard methods from the UK and elsewhere

Box 1.2 Sampling protocols

(General Guidance Note 4 (1991), Department of the Environment)
Emission testing requirements should normally include a sampling protocol, so as to ensure that emissions are

properly characterized, representatively sampled and correctly assessed.

The exact nature of the protocol will depend on both the process to be assessed and the specified emission sampling and testing methods. Any sampling protocol should specify the following:

a) The **range of pollutants** to be measured, the **standard method** and the **degree of accuracy** required
b) The **process operating conditions** at the time of sampling (e.g. process load and flow rate)
c) The **minimum number of tests** and **overall sampling time** necessary to obtain representative samples.
d) The **time periods for individual sampling** and testing in order to detect any anticipated variation in emissions.
e) The appropriate **calibration** of measurement techniques (including laboratory quality control)
f) The format for **recording, reporting** and **statistical treatment** and **assessment** of results.

(HMIP, 1993b) while 'M3' and 'M4' survey the international standards organizations and draw together methods relevant to monitoring in processes covered by Integrated Pollution Control (HMIP, 1995a,b).

Several standard methods are available for smoke and dust emissions. That normally required for extractive sampling is BS 3405, although there has been a more stringent, rarely used BS 893. The latter is now obsolete following the addition of BS 6069 (1992) which is the British equivalent of an international ISO method. The visual assessment of smoke emissions is covered by BS 2811 and 2472 (see Chapter 4) although the practitioners of this art are a disappearing breed! The standard for opacimeters is rather old and does not reflect modern technological capability. Methods of particle size analysis are covered by BS 3406 but none of the methods outlined relates to flue gases.

Some of the test methods for gases described in BS 1756 'Sampling and analysis of flue gases' now look rather dated. A number of these are covered in Chapter 6. Part 4 of the new BS 6069 covers stationary source emissions, but to date the only gas analysis method detailed is for SO_2. Another section of this standard discusses performance characteristics for automated methods of measuring SO_2. Older standards covering automated methods of analysis are so dated as to be of little relevance. Methods for the preparation of calibration gases, which are discussed in Chapter 13, are detailed in the various sections of BS 4559. All these are listed in Appendix A.

It is likely that in the future the UK will, wherever possible, adopt international (ISO) or European (CEN) standard methods rather than going it alone in the production of a large number of new national standards. The interactions between the various standards organizations are described in Technical Guidance Note M3 (HMIP, 1995a). This lists the working groups preparing future standards on stationary source emissions under CEN Technical Committee TC264, the BSI subcommittee EPC/35/1 and ISO TC146 subcommittee SC1.

The Institution of Environmental Health Officers in conjunction with Warren Spring Laboratory (now National Environmental Technology Centre) have produced a three volume manual for EHOs dealing with Part B processes. This is likely to become the reference work for much emissions work in the UK in future. Details are available from Chadwick House Group Ltd, 16 Great Guildford St, London SE1 0ES.

For some pollutant emissions it may be possible to use the same analytical techniques as are used for their determination in the occupational environment (although the sample collection method will differ). A wide range of substances are covered in the MDHS series of booklets, *Methods for the Determination of Hazardous Substances in Air*, published by Health and Safety Executive. Some of these are now appearing as British Standards in section 3 of BS 6069 and these are listed in Appendix A.

1.3.3 German standard methods

Technical guidelines on pollution control including methods of testing emissions are prepared by the Association of German Engineers (VDI Verein Deutscher Ingenieure). A list of VDI methods is included in Appendix B.

1.3.4 US standard methods

Performance standards and emissions limits in the USA are always specified in conjunction with a specified method of measurement. The EPA approved methods for particles, gases and the initial group of Hazardous Air Pollutants are listed in Appendix C. There are also a number of ASTM standards which are relevant to emissions monitoring.

1.3.5 Instrument performance specifications

There has been very little information available in the UK on specifications that emissions monitoring equipment must meet, apart from the 1993 section of BS 6069 dealing with SO_2 analysers mentioned above. The two old British Standards (BS 3048, 1958 and BS 5849, 1980) on continuous monitoring instruments have never been updated. The information in the Process Guidance Notes tends to be of a rather general nature and often amounts to little more than that the controlling authorities (Environment Agency or Local Authorities) must be satisfied that the instrumentation and procedures used are satisfactory. However, following discussions between various parties in the UK, a proposal for a Monitoring Certification Scheme was drawn up in 1996 and this scheme will set the standards of performance which instruments should achieve. The report (Environment Agency, 1996) covers the proposed certification scheme, performance standards, and laboratory and field test procedures for continuous emission monitoring instruments.

In Germany, not only are the test methods involving continuous emissions monitors specified but an instrument approval system also operates. Guidelines for the specifications which must be met and the instrument testing procedures are set by the Federal Ministry of the Environment. The actual testing and approval is carried out by regional Technical Inspection Offices (TÜV, Technische Überwachungs-Verein). Manufacturers must provide three instruments, together with their own specifications. Both laboratory tests and a 3-month site trial are carried out before approval can be given. A list of approved instruments for specific application is published.

In the USA, performance specifications for continuous emissions monitoring systems and details of quality control procedures are published in addition to the approved methods. A list is given in Appendix C and further discussion of the procedures for calibration of continuous emissions monitoring systems are discussed in Chapter 12. Since, under the US allowance system for SO_2, the accuracy of the measurement of the total emissions has direct financial implications, they have an added incentive for ensuring that calibration and other instrument checks are correctly carried out. The US EPA also operate an instrument approval system and instrument manufacturers must have their individual models tested and certified.

REFERENCES

HMIP (1996) The Chemical Release Inventory 1994. H.M. Inspectorate of Pollution, Department of the Environment, HMSO, London.

Murley, L. (ed.) (1995) *Clean Air Around the World*, 3rd edn, International Union of Air Pollution Prevention and Environmental Protection Associations, Brighton.

Boehmer-Christiansen, S. and Skea, J. (1991) *Acid Politics: Environmental and Energy Policies in Britain and Germany*, Belhaven Press, London.

Federal Ministry for the Environment, Nature Conservation and Nuclear Safety (1988) *Air Pollution Control Manual of Continuous Emission Monitoring. Regulations and Procedures for Emission Monitoring*. 2nd revised edition.

Code of Federal Regulations, Title 40 Protection of the Environment, chapter 1 Environmental Protection Agency contains all the regulations relating to air programs (Parts 1–99). (Details of these can be obtained from the address in Appendix C.)

HMIP Technical Guidance Notes (Monitoring) M1 (1993a) Sampling facility requirements for the monitoring of particulates in gaseous releases to the atmosphere. HMSO, London.

HMIP Technical Guidance Notes (Monitoring) M2 (1993b) Monitoring emissions of pollutants at source. HMSO, London.

HMIP Technical Guidance Notes (Monitoring) M3 (1995a) Standards for IPC Monitoring: Part 1 – Standards organizations and the measurement structure.

HMIP Technical Guidance Notes (Monitoring) M4 (1995b) Standards for IPC Monitoring: Part 2 – Standards in support of IPC monitoring.

Environment Agency (1996) The Establishment of a Monitoring Certification Scheme – Proposed Certification Scheme, Performance Standards, and Laboratory and Field Test Procedures for Continuous Emission Monitoring Instruments. Report available from National Centre for Regulatory Monitoring, Lancaster.

2

Gas composition calculations

Bernard M. Gibbs

2.1 UNITS

2.1.1 Volume and mass basis units

Air pollutant emissions may be measured either in volume-based units or mass-based units:

Volume basis		Mass basis
%	1 in 100	$g\ m^{-3}$
ppm	1 in 10^6	$mg\ m^{-3}$
ppb	1 in 10^9	$\mu g\ m^{-3}$
ppt	1 in 10^{12}	$ng\ m^{-3}$

It is important to recognize that the concentration value measured in volume units is independent of temperature and pressure since it is effectively a ratio (m^3 per m^3). Concentrations expressed in mass units vary with temperature and pressure and are normally quoted at NTP (0°C and 1 atm.). This is sometimes written $\mu g\ Nm^{-3}$ (microgram per normal meter cubed). Assuming ideal gas behaviour, the proportion of a pollutant by volume is the same as that by number of moles – 1 ppm corresponds to a mole fraction of 10^{-6}.

2.1.2 Converting ppm emissions to mass emissions

Frequently emission standards are expressed in mg/m^3 or $\mu g/m^3$ and ppm have to be converted to mass emissions for comparative purposes. The steps involved are as follows.

$$[\text{ppm}] = \frac{\text{m}^3}{\text{m}^3} \times 10^{-6}$$

but at standard conditions (273 K, 101.33 $kN\ m^{-2}$ or 1 atm),

$$1\ \text{m}^3 \equiv [1/22.4]\ \text{k mol} \equiv [\text{MW}/22.4]\ \text{kg}$$

where MW = molecular wt of pollutant (kg/kmol).

[N.B. The term Relative Molecular Mass (RMM) is replacing the older term molecular weight.]

$\therefore$[ppm] = [MW/22.4] × 10^{-6} kg/m^3
but 1 kg = 10^6 mg = 10^9 μg
$\therefore$[ppm] = [MW/22.4] mg/m^3 or [MW/22.4] × 10^3 μg/m^3

Thus, if the emission e is, expressed in ppm, then the mass emissions, E_m (mg/m^3) and E_μ (μg/m^3) are given by

E_m = e [MW/22.4] and
$E\mu$ = e [MW/22.4] × 10^3

Conversion factors for some of the common pollutant gases are given in Table 2.1.

Table 2.1 Conversion from volume to mass basis

Gas	ppm to mg/Nm^3 multiply by	mg/Nm^3 to ppm multiply by
CO	1.25	0.800
CO_2	1.96	0.510
NO	1.34	0.747
NO_2	2.05	0.487
SO_2	2.86	0.350
HCl	1.63	0.613

Allowing for different temperatures or pressures

The above formulae apply to standard conditions (273K, 1 atm), if the emission is required at any other temp and pressure then application of the gas laws yields:

$$[E]_2 = [E]_1 (T_1/T_2)(P_2/P_1)$$

Example

A plant has an emission of 1250 ppm HCl; determine the emission in mg/m^3 (NTP).

(i) Calculation of mass emission at NTP
MW, HCl = 36.5
E_m = [36.5/22.4] × 1250
= 2037 mg/m^3

(ii) If we want E_m* at another T and P, say P = 2 atm, T = 500 K, then
$[E_m]^*$ = $E_m \times T_1/T_2 \times P_2/P_1$
= 2037 × [273/500] × 2
= 2224 mg/m^3

2.1.3 Thermal basis units

In the USA, some emission limits are given in terms of the thermal input of the plant in millions of Btu. The conversion from volume basis units to a thermal basis depends on the specific fuel and the standard reference conditions (% O_2). Table 2.2 gives the factors.

Table 2.2 Conversion from mass per unit volume to thermal basis at 6% O_2 for solid fuels and 3% O_2 for gaseous or liquid fuels

Fuel	mg/Nm^3 to g/GJ multiply by	mg/Nm^3 to lb/10^6Btu multiply by
Anthracite	0.382	8.89×10^{-4}
Bituminous coal	0.370 (0.350)	8.61×10^{-4} (8.14×10^{-4})
Lignite	0.373	8.68×10^{-4}
Liquid fossil fuels	0.289 (0.280)	6.73×10^{-4} (6.51×10^{-4})
Gaseous fossil fuels	0.274 (0.270)	6.38×10^{-4} (6.28×10^{-4})

NB. Conversion factors based on data given in the US 'Standards of performance for fossil-fuel-fired steam generators' 40 CFR Section 60.45. Figures in brackets are based on a 1986 IEA paper.

2.2 WET AND DRY GAS

Moisture is frequently present in gaseous emissions whether it arises from the water content of the fuel or feedstock, moisture present in the air, combustion of hydrocarbons or from the use of aqueous solutions in the process. Emissions limits are normally expressed on a *dry gas basis,* which means that variable moisture does not affect the assessment of the emissions. If the percentage water vapour content is $[H_2O]$ then the relationship between the wet gas and dry gas emissions is

$$E_{dry} = E_{wet} \frac{100}{(100 - [H_2O])}$$

As can be seen, the dry gas reading is always greater than the wet gas reading.

2.3 COMBUSTION FUNDAMENTALS

2.3.1 Combustion air

The main components of dry air are nitrogen (78.1%), oxygen (20.95%), argon (0.9%) and carbon dioxide (0.035%, 350 ppm). This excludes the water vapour, which can be up to about 2%. For all practical purposes we can neglect the CO_2, and group the argon in with nitrogen. Thus, dry air can – for calculation purposes – be assumed to contain 21.0% by vol. of oxygen and 79.0% by vol. of nitrogen. The volume or molar ratio of N_2/O_2 is 79/21 or 3.76 and this will be a useful number in what follows.

2.3.2 Theoretical or stoichiometric air requirements (Vs)

The combustible matter in fuels and wastes is usually a complex mixture of organic compounds whose principal components are C, H and O. It is essential that combustible matter is burnt completely to produce CO_2 and H_2O as the products, in accordance with the combustion reactions, summarized in Table 2.3.

The equations in Table 2.3 define precisely the quantity of oxygen theoretically required to ensure complete combustion, irrespective of both the mechanism of the combustion reactions and the way in which C, H, O, etc. are combined in the fuel.

As the composition of air is known, the *theoretical* or *stoichiometric air* can easily be calculated by multiplying the stoichiometric oxygen by the molar ratio of N_2/O_2, i.e. 3.76.

Table 2.3 Combustion equations

$C + O_2 \rightarrow CO_2$
$C + \frac{1}{2}O_2 \rightarrow CO$ (incomplete combustion)
$CO + \frac{1}{2}O_2 \rightarrow CO_2$
$S + O_2 \rightarrow SO_2$
$H_2 + \frac{1}{2}O_2 \rightarrow H_2O$
$CH_4 + 2O_2 \rightarrow CO_2 + 2H_2O$
$C_2H_6 + 7/2O_2 \rightarrow 2CO_2 + 3H_2O$
$C_3H_8 + 5O_2 \rightarrow 3CO_2 + 4H_2O$
$C_2H_4 + 3O_2 \rightarrow 2CO_2 + 2H_2O$
$C_3H_6 + 9/2O_2 \rightarrow 3CO_2 + 3H_2O$
$C_2H_2 + 3/2O_2 \rightarrow 2CO_2 + H_2O$
$C_6H_6 + 15/2O_2 \rightarrow 6CO_2 + 3H_2O$

2.3.3 Excess air (ϵ)

In almost all practical combustion systems it is necessary to use more than the stoichiometric amount of air, to ensure complete combustion. This additional air is referred to as *excess air*, and is denoted as a percentage or fraction over the stoichiometric amount.

The maximum concentrations of carbon dioxide and water vapour in the flue gases are found close to the stoichiometric point. As more and more excess air is supplied, the CO_2 and H_2O concentrations decrease simply due to dilution while the O_2 level of course rises. Figure 2.1 shows the trends for three different fuels.

If less than stoichiometric air is supplied there will be insufficient oxygen to convert all of the carbon in the fuel to CO_2 and all of the hydrogen

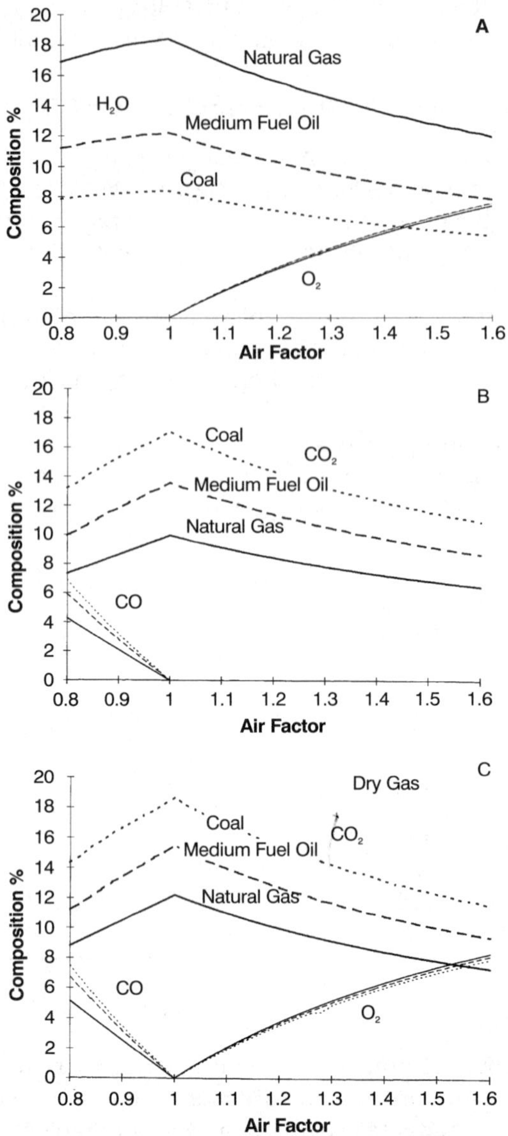

Fig. 2.1 Theoretical equilibrium calculations of combustion gas composition for different fuels as a function of 'air factor' defined such that for stoichiometric combustion the air factor = 1.0, while 20% excess air corresponds to air factor 1.2, etc. Note that under sub-stoichiometric conditions (air factor <1) the values are sensitive to the temperature at which the equilibria between the various species (H_2O, O_2, CO, CO_2, H_2) is determined. In these figures 1400K has been assumed. The curves relate to a typical bituminous coal in 'as received' state, i.e. including a significant amount of moisture. (A) H_2O and O_2 wet gas; (B) CO_2 and CO wet gas; (C) CO_2, CO and O_2 dry gas.

in the fuel to H_2O. Significant amounts of CO and hydrogen may then be formed together with elevated levels of unburned or partially burned fuel. If the temperature is high enough and the residence time long enough the composition would tend to the equilibrium position of the water gas reaction:

$$CO + H_2O = CO_2 + H_2$$

The theoretical equilibrium levels of CO are illustrated in Fig. 2.1 but in practice the actual composition will be determined by a combination of kinetic and equilibrium factors related to the rate of cooling and sophisticated reaction rate models are necessary to estimate the composition.

For most purposes these detailed equilibrium or kinetic calculations are unnecessary and calculations of product gas composition and air requirements can be obtained with adequate accuracy using simple estimates based on stoichiometry alone. The next two sections describe these methods with examples.

2.4 THE 'MOLE' METHOD OF CALCULATION FOR GASES

The air requirements, volume and composition of the products are most easily determined by working in kmols. 1 kmol occupies 22.41 m^3 at 273K, 1 atm. The simplest case is that of a fuel gas, because 1 mol of gas occupies the same volume as 1 mol of another gas (at the same temperature and pressure) and in molar terms, excess air, ϵ, is defined by

$$\epsilon = \frac{V_a - V_s}{V_s}$$

where V_a = actual air, kmol or m^3, and V_S = stoichiometric air, kmol or m^3.

For solid fuels, a conversion to kmol is required (see later).

Combustion calculations may take two routes.

1. Via the fuel analysis.
2. Utilizing the flue gas analysis.

Calculations making use of the flue gas analysis are discussed in section 2.6 below. Calculations based on the fuel analysis provide:

- combustion air and waste gas flows;
- heat released to be estimated; and
- emissions to be predicted – gaseous and ash produced.

Note that the waste gas flow is needed to estimate the efflux velocity at the chimney top, the plume rise and dispersion of the gases in the atmosphere, the pressure drops through the plant, etc.

Example: gaseous fuel calculation
The composition of a fuel gas is given as methane (CH_4) 93.0%, propane (C_3H_8) 2.0%, hydrogen sulphide (H_2S) 2.0%, nitrogen (N_2) 3.0%. Calculate, for combustion with 30% excess air,
(i) the air and waste gas volume per m^3 of gas combusted; and
(ii) the SO_2 emission in ppm on a wet and dry basis.

N.B. The composition of gaseous fuels is the volumetric or molar composition (since these are equivalent). Hence, the calculation uses the stoichiometric combustion reaction for each component to give

- the oxygen required; and
- the products produced

in units of volume (m^3) or kmol.

Step 1. Write down combustion reactions

$CH_4 + 2O_2 \rightarrow CO_2 + 2H_2O$
$C_3H_8 + 5O_2 \rightarrow 3CO_2 + 4H_2O$
$H_2S + 1.5O_2 \rightarrow SO_2 + H_2O$

Step 2. Tabular calculation

Species	% v/v or $m^3/100\ m^3$	O_2 required	Product volumes $V_p\ m^3$			
			CO_2	SO_2	H_2O	N_2
CH_4	93.0	186.0	93.0	–	186.0	–
C_3H_8	2.0	10.0	6.0	–	8.0	–
H_2S	2.0	3.0	–	2.0	2.0	–
N_2	3.0	–	–	–	–	3.0
stoichiometric O_2		199.0	–	–	–	–
stoichiometric N_2	$= 3.76\times O_2$	748.2	–	–	–	748.2
Volume totals		947.2	99.0	2.0	196.0	751.2
Component		air	CO_2	SO_2	H_2O	N_2

First, gather the relevant total volumes for stoichiometric combustion:
Stoichiometric air:

$$V_s = 947.2\ m^3/100\ m^3 = 9.47\ m^3/m^3$$

Stoichiometric waste gas, wet,

$$W_{sw} = \Sigma V_p = 1048.2\ m^3/100\ m^3$$
$$= 10.5\ m^3/m^3$$

Stoichiometric waste gas, dry:

$$W_{sd} = \Sigma V_p - V_{H_2O} = 10.5 - 196.0/100$$
$$= 8.5\ m^3/m^3$$

Then we can calculate the required parameters with excess air present:

(i) Actual air required, V_a

$V_a = V_s(1+\epsilon)$, ϵ = fraction of excess air

but $V_s = 9.47\ m^3/m^3$ and $\epsilon = 0.30$

$\therefore\ V_a = 9.47 \times 1.3$

$= 12.3\ m^3/m^3$ gas burnt

(ii) Actual waste gas flows, wet and dry, W_{aw}, W_{ad}

$W_{aw} = W_{sw} + \epsilon V_s$

but $W_{sw} = 10.5$, $V_s = 9.47$, $\epsilon = 0.3$

$\therefore\ W_{aw} = 13.34\ m^3/m^3$ gas burnt

and $W_{ad} = W_{sd} + \epsilon V_s$

$= 8.5 + 0.3 \times 9.47$

$= 11.34\ m^3/m^3$ gas burnt

(iii) SO_2 emissions, wet and dry, [SO2]w and [SO2]dry
The concentration (as a volume ratio) is the pollutant gas volume divided by the total waste gas volume. On a wet basis:

$$[SO_2]_w = V_{SO_2}/W_{aw} = (2.0/100) \div 13.34$$
$$= 1.499 \times 10^{-3}\ m^3/m^3$$
$$[ppm\ SO_2]_w = [SO_2]_w \times 10^6$$
$$= 1499\ ppm$$

And on a dry basis

$$[SO_2]_d = V_{SO_2}/W_{dw} = (2.0/100) \div 11.34$$
$$= 1.763 \times 10^{-3}\ m^3/m^3$$
$$= 1763\ ppm$$

2.5 COMBUSTION AND EMISSIONS CALCULATIONS FOR SOLID AND LIQUID FUELS OR WASTES

As before, the starting point is the analysis of the fuel or waste. However, the molar composition is unknown and the analysis of the fuel is given in terms of the elemental composition or ultimate analysis which determines the weight fraction of each component element in the fuel, i.e. C, H, O, S, N, etc. In addition, the proximate analysis yields the fractions of ash and moisture in the fuel.

Calculation procedure

1. A molar/stoichiometric procedure is again used, but before this can be undertaken, the elemental composition must first be converted into a molar composition. This is achieved by dividing each component by its appropriate molecular weight.
2. The presence of oxygen in the fuel is accounted for by taking it from the total oxygen required by all the other combustible components.

This is conveniently carried out by assigning a negative sign to the molar oxygen content of the fuel in the Tabular calculation.

3. Some components of the fuel/waste may not participate in the combustion and can: (i) pass through the combustor chemically unchanged – this usually applies to ash, water; or (ii) be released as a decomposition product, combined with another component, e.g.

   Chlorine – released as HCl
   Fluorine – released as HF

The combustion calculation therefore needs to take account of combustible components that are lost due to decomposition. This can be undertaken at the 'molar' stage.

The following example illustrates the calculation procedure for a waste with decomposition products.

Example

A waste of the following composition (by wt): C 50%, H 10%, O 20%, N 2.0%, Cl 3.5%, F 0.5%, moisture 10.0% and ash 4.0%, is burnt with 100% excess air. Assuming complete combustion, determine

1. The combustion air flow-rate and waste gas flow for combustion of 1 tonne/h of waste.
2. The emission of HCl and HF from the dry flue gas.

Reactions

$$C + O_2 \rightarrow CO_2$$
$$H_2 + \tfrac{1}{2}O_2 \rightarrow H_2O$$
$$\left.\begin{array}{l} H_2 + Cl_2 \rightarrow 2\,HCl \\ H_2 + F_2 \rightarrow 2\,HF \end{array}\right\} \text{equivalent to decomposition}$$

Tabular calculation

Solid/liquid fuel/waste
Stoichiometric combustion

Element	kg/ 100 kg	kmol/ 100 kg	crtd* k mol	O_2 rqd†	Product kmol CO_2	H_2O	N_2	HCl	HF
C	50	4.17		4.17	4.17				
H as H_2	10	5.0	4.94	2.47		4.94			
O as O_2	20	0.63		–0.63					
N as N_2	2.0	0.07	↑	–			0.07		
Cl as Cl_2	3.5	0.05 → →	↑		–			0.1	
F as F_2	0.5	0.013→→	↑		–				0.026
H_2O	10.0	0.56		–		0.56			
Ash	4.0	–							
		stoichiometric O_2		6.01					
[O_2 (× 3.76)]		stoichiometric N_2		22.60			22.60		
Volume totals		stoichiometric air		28.61	4.17	5.5	22.67	0.1	0.026
Components				air	CO_2	H_2O	N_2	HC1	HF

*corrected †required

Stoichiometric waste gas, wet Σ_{wet} = 32.47 kmol/100 kg fuel
Stoichiometric waste gas, dry Σ_{dry} = 26.97 kmol/100 kg fuel

We can now calculate the required parameters with excess air

(i) Combustion air per 100 kg

$V_a = V_s(1 + \epsilon)$.
V_s = 28.61 kmol/100 kg (Table)
ϵ = 1.0
$\therefore\ V_a$ = 28.61 (1 + 1) = 57.22 kmol/100 kg
$\therefore$ air/1000 kg = 10 × V_a = 572.2 kmol/tonne
= 572.2 × 22.4 m^3
= 12817 m^3 per tonne (273 K, 1 atm)

(ii) Waste gas flow, wet, per 100 kg

$W_{aw} = [\Sigma_{wet} + \epsilon V_s]$
= [32.47 + 28.61] kmol/100 kg
= 61.08 mol
WG/1000 kg = 10 × 61.08 = 610.8 kmol
= 610.8 × 22.4 m^3 (at 273K, 1 atm)
= $610.8 \times 22.4 \times \frac{500}{273}$ m^3 (at 500K, 1 atm)
Waste gas = 25058 m^3 (500K, 1 atm) per tonne

(iii) Emission of HCl and HF – dry basis

Since $W_{ad} = \Sigma_{dry} + \epsilon V_s$ = 55.58
Emission of HCl = V_{HCl}/W_{ad} = 0.10/55.58 m^3/m^3
= 1799 ppm.
Emission of HF = 0.026/55.58 = 0.468 × 10^{-3} m^3/m^3
= 468 ppm

2.6 CALCULATIONS BASED ON FLUE GAS ANALYSIS

These enable

1. plant performance to be assessed
2. excess air level to be obtained
3. emissions to be compared with standards set at reference conditions of moisture and oxygen.

In addition, in many cases they may be the only means of monitoring the plant, as a fuel analysis may not be possible.

2.6.1 Excess air level (ϵ)

Measurements available

1. ϵ can be obtained by measuring the O_2 or CO_2 content of the flue gas. A simple formula can then be applied to obtain ϵ.
2. Wet or dry analysis can be used depending on the instruments available.

For O_2 analysis

$$[O_2 \text{ in flue gas}] = \frac{[O_2 \text{ in excess air}]}{[\text{vol. of flue gas}]}$$

$$= \frac{[\epsilon.V_s] \times 21}{[W_s + \epsilon.V_s]}$$

Re-arranging

$$\epsilon = \frac{[O_2]}{[21 - O_2]} \times \frac{W_s}{V_s}$$

when $W_s \approx V_s$, then

$$\epsilon \approx \frac{[O_2]}{[21 - O_2]}$$

For CO_2 analysis – usually dry basis

$$[CO_2 \text{ in flue gas}] = \frac{[\text{vol. of stoich. } CO_2]}{[\text{vol. of flue gas}]}$$

$$\therefore [CO_2]_a = \frac{W_{sd}\cdot[CO_2]_s}{[W_{sd} + \epsilon V_s]}$$

Re-arranging:

$$\epsilon = \frac{W_{sd}}{V_s}\left[\frac{[CO_2]_s}{[CO_2]_a} - 1\right]$$

$$\approx \left[\frac{[CO_2]_s}{[CO]_a} - 1\right]$$

The accuracy of the assumption that $W_s = V_s$ on either a wet or dry basis is indicated in the table below:

Stoichiometric waste gas(W_{sw} or W_{sd})/air (V_s) ratios	Wet	Dry
Methane	1.105	0.895
Heavy oil	1.060	0.940
Coal (bituminous)	1.052	0.975
Waste	1.153	0.95

2.6.2 Normalizing emissions to reference conditions

Measured emissions are 'normalized' to a standardized oxygen level and water vapour content (usually dry) in order to overcome dilution effects caused by differences in excess air levels or moisture. Figure 2.1 illustrates the effects for fuel combustion, but the same principle of course applies to any process where air or moisture may be present. The reference oxygen levels are typically

- 3% for gas or liquid fuels
- 6% for solid fuels
- 11% for incineration processes

The correction of an emission to a standard requires a knowledge of the excess air or O_2 level. A simple formula in then applied, which is derived below.

The emission, e (ppm or vol/vol), at any excess air level is given by

$$e = \frac{[\text{vol. of pollutant}]}{[\text{vol. of flue gas}]}$$

At an excess air level, ϵ_1, then the emission e_1 is

$$e_1 = \frac{V_p}{[W_s + \epsilon_1 V_s]}$$

where V_p = volume of pollutant generated, and for an excess air level, ϵ_2,

$$e_2 = \frac{V_p}{[W_s + \epsilon_2 V_s]}$$

equating the emission V_p

$$e_2 = e_1 \frac{[1 + \epsilon_1 V_s/W_s]}{[1 + \epsilon_2 V_s/W_s]} \approx \frac{e_1 [1 + \epsilon_1]}{[1 + \epsilon_2]}$$

Using our previous equation for the excess air expressed in terms of the O_2 concentration, leads to the exact expression

$$e_2 = \left[\frac{21-[O_2]_2}{21-[O_2]_1}\right] e_1$$

Example

A plant has an emission of 1000 ppm HCl at an oxygen concentration of 13% (wet basis) and water vapour 8%; determine the emission in mg/m^3 (NTP) at 11% oxygen (dry basis).

(i) Correct concentrations to dry basis

$$[HCl]_{dry} = [HCl]_{wet}\,100/(100 - [H_2O])$$
$$= 1000 \times 100/92$$
$$= 1087 \text{ ppm}$$

Similarly,

$$[O_2]_{dry} = 13 \times 100/92$$
$$= 14.13\%$$

(ii) Correct emission to 11% O_2

$$[HCl]_{11} = \frac{[21 - 11]}{[21 - 14.13]} \times 1087 = 1582 \text{ ppm}$$

(iii) Mass emission at NTP

$$\text{MW, HCl} = 36.5$$
$$E_m = [36.5/22.4] \times 1582$$
$$= 2578 \text{ mg/m}^3$$

Frequent reference to normalization will be made elsewhere, especially in the chapter dealing with data logging where the ability to undertake the necessary calculations automatically and in real time is discussed.

3

Particulate emissions by extractive sampling

Andrew G. Clarke and George Bartle

3.1 INTRODUCTION

In contrast to the measurement of gaseous emissions which can be routinely undertaken with an accuracy of a few percent, the measurement of particulate emissions is far more difficult. This arises primarily from the maldistribution of particle concentration within the duct or chimney coupled to the non-uniformity of the gas velocity. It is also partly due to sampling difficulties. Variations in the particle concentration from point to point within a duct or flue may arise for several reasons. These include incomplete mixing downstream of a junction if the two gas streams bear different particle loads, or incomplete mixing downstream of a point where air has leaked into the process gas stream. Since the air is likely to be relatively cool this problem should be detected during the preliminary probing of the gas temperature which is a normal part of the testing procedure. It may then be necessary to seal the leak or to move the sampling position. The other factors which upset the uniformity of the particle distribution are sedimentation and inertial effects when the gas passes round bends or obstacles to the flow. These factors are discussed in the next section.

The basic requirement of all extractive sampling techniques is that the sample of gas taken into the measurement system should be *representative* of the bulk gas stream in the flue. Non-representative sampling can arise because of non-uniform conditions in the flue coupled with the use of an inadequate number of sampling points to reflect these conditions. Alternatively, it can arise because the gas composition is somehow altered in the process of sampling (particulate concentration changed). For these reasons very precise guidelines for particle sampling are required and these are given in the various standard methods.

The normal requirement in the UK is to test according to BS 3405 (1983) which quotes an accuracy of ±25%. There is a somewhat more precise sampling method described in BS 893 which states that the overall accuracy 'is unlikely to be better than ±10% under ideal conditions, i.e. test rig conditions'. The main difference between the two standards is that the latter requires a larger number of sampling points. The more recently published BS 6069, Section 4.3 (1992) is equivalent to the ISO standard 9096 (1992). It remains to be seen whether this will in future be the method recommended by the authorities in the UK and for this reason we shall concentrate on BS 3405. The main provisions of this standard are given in Appendix D. In the USA, the EPA Method 5 (Appendix C) is most commonly used, although comments will also be made about Method 17. German methods are given in various parts of the VDI guideline number 2066 (Appendix B). In the following section the basic particle properties and the implications for particle sampling are considered. The hardware for particle sampling is described in section 3.3 and the procedures and practical aspects of sampling in section 3.4.

3.2 BASIC PRINCIPLES

3.2.1 Particle properties

It will be helpful to clarify the terminology used in the UK in relation to gas-borne particles. *Grit* was defined in the Clean Air (Emission of Grit and Dust from Furnaces) Regulations 1971 as particles exceeding 76 micrometres (μm) in diameter. The size relates to the aperture of a 200 BS mesh sieve. *Dust* particles are below 76 μm but above about 1 μm. The lower limit is not precisely defined. Sub-micrometre particles are referred to as *fume*. Fumes tend to be generated in processes where there is gas-to-particle conversion either by condensation or by chemical reaction, such as metallurgical fumes and soot. BS 3405 relates specifically to the measurement of grit and dust, although fume would also be measured by this standard, providing that an efficient filtration system is used.

Sedimentation under gravity of larger particles may result in higher concentrations accumulating along the bottom of horizontal ducts. The terminal velocity U of spheres of diameter d and density ρ is given by the relationship

$$U = \rho d^2 K_s g / 18\eta$$

where K_s is a slip correction which is most significant for small particles, g is the acceleration due to gravity and η is the gas viscosity. In this equation the density of the gas is assumed to be negligible compared with the particle density.

Real particles are generally not spherical and the equation should then be written in terms of the effective or *Stokes diameter*. The Stokes diameter of a particle is defined as the diameter of a sphere having the same bulk density and the same terminal settling velocity as the particle. BS 3405 specifically refers to the size of dust in terms of the Stokes diameter of the particles.

Typical terminal velocities are given in Table 3.1. As can be seen the velocity at the lower limit of the 'grit ' range is of the order of 20 cm/s and exceeds 1 m/s above 200 μm when sedimentation problems will be severe.

In industrial plant the duct diameters and velocities are such that the Reynold's Number is high and one is dealing with turbulent rather than laminar flow. In ideal circumstances the mean velocity profile across the duct should be fairly flat except for the decreased values in the boundary layer near the walls (Fig. 3.1, plug flow). The likely profiles downstream of a bend, a partially closed damper and an axial fan are also illustrated in the figure. With the axial fan there is the additional problem that the flow will be swirling not parallel to the duct axis. For a constant dust concentration the mass flux across the sampling plane will vary from point to point in proportion to the velocity. However, the dust concentration is generally not constant.

When the gas flow deviates from a straight line the finest particles will follow the gas streamlines, but the larger particles – due to their greater momentum – will tend to continue along their original path. The angle between the particle track and the flow streamline will depend on the balance between the inertial force and the viscous drag force and this is expressed in terms of a dimensionless parameter called the *Stokes number,* Stk:

$$\mathrm{Stk} = UV/gr = \rho d^2 K_s V / 18\eta r$$

Table 3.1 Terminal velocities (m/s) of spheres of different sizes and density 2000 kg/m³ in air (Hawksley *et al.*, 1977)

	Temperature	
Diameter (μm)	15°C	200°C
5	1.5×10^{-3}	1.1×10^{-3}
10	6.1×10^{-3}	4.2×10^{-3}
20	2.4×10^{-2}	1.7×10^{-2}
50	1.4×10^{-1}	1.0×10^{-1}
70	2.6×10^{-1}	2.0×10^{-1}
100	4.6×10^{-1}	3.7×10^{-1}
250	1.2	1.1
500	3.2	3.3

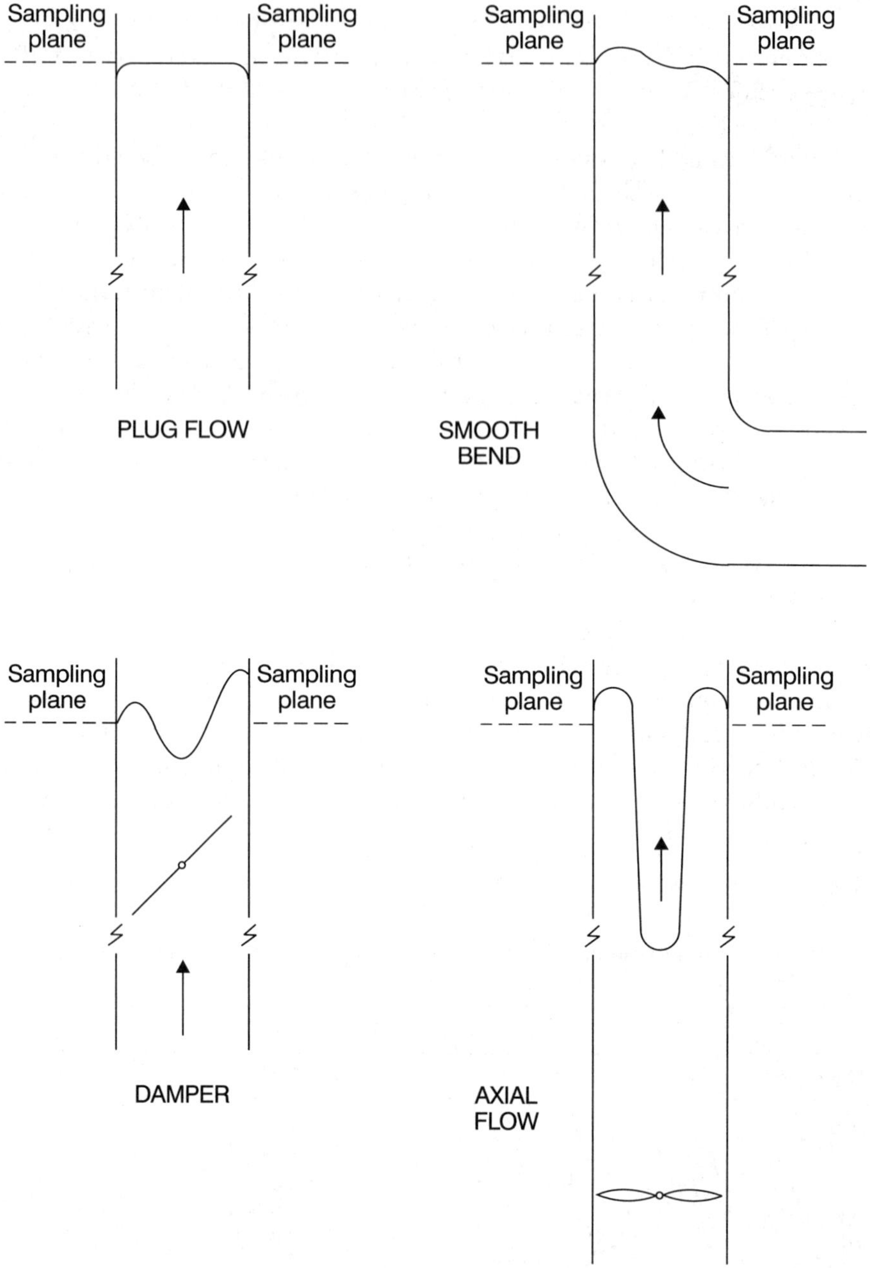

Fig. 3.1 Typical turbulent flow velocity profiles in ducts.

where V is the gas velocity and r is the radius of the bend round which the gas is passing. This parameter appears frequently in the design of particle collection devices such as impactors and cyclones.

The extent of the deviations from uniformity of both the gas velocity and the particle emissions rate which may be experienced in practice are illustrated in Fig. 3.2 taken from Hawksley *et al.* (1977). The sampling plane is shaded and the results are expressed relative to the averages of the two parameters for the whole duct. The emissions rate is the flux of particles per unit area per unit time. Figure 3.2A illustrates the effect of a bend. On the outside of the bend the gas velocity is enhanced by 20% compared with the average, while on the inside of the bend the velocity is 20% lower than average. However, the emissions are even more non-uniform with 50% enhancement on the outside of the bend and up to 60% reduction on the inside of the bend. Figure 3.2B illustrates similar effects downstream of a fan. There are greatly enhanced velocities on the bottom of the duct and an almost stagnant region in the top left hand corner. The flux of particles varies from 0.1 to 2.2 times the average value for the duct.

The conclusions of this discussion of particle properties are firstly that choice of an appropriate sampling position is vital, and secondly that multipoint sampling must be used. Only for the smallest ducts or chimneys where it is impractical to use the standard methods might single point sampling be necessary.

3.2.2 Location of sampling position

Due to the effects of sedimentation there will inevitably be a layer of particulate material on the bottom horizontal surface of a duct and, from this, resuspension may take place causing very variable concentrations. To minimize such errors a length of vertical duct or the chimney itself is preferable to a section of horizontal duct. This may not in practice be convenient, since access to an appropriate level on the chimney may be difficult and the sampling position will be exposed to the elements rather than being under cover.

The problem of inertial effects requires that locations close to features that disturb the flow must be avoided. 'The sampling position shall be downstream of any particulate arrestment plant. It shall be located in a straight length of flue, as far downstream as practicable from any obstruction (e.g. bend, fan or partially closed damper) that may cause a disturbance and produce a change in the direction of gas flow.' (BS 3405, 1983). The standard, recognizing that life may not always be simple does give more specific guidance. If the diameter of the flue is D, the sampling position should be not less than the following distances from flow disturbing features:

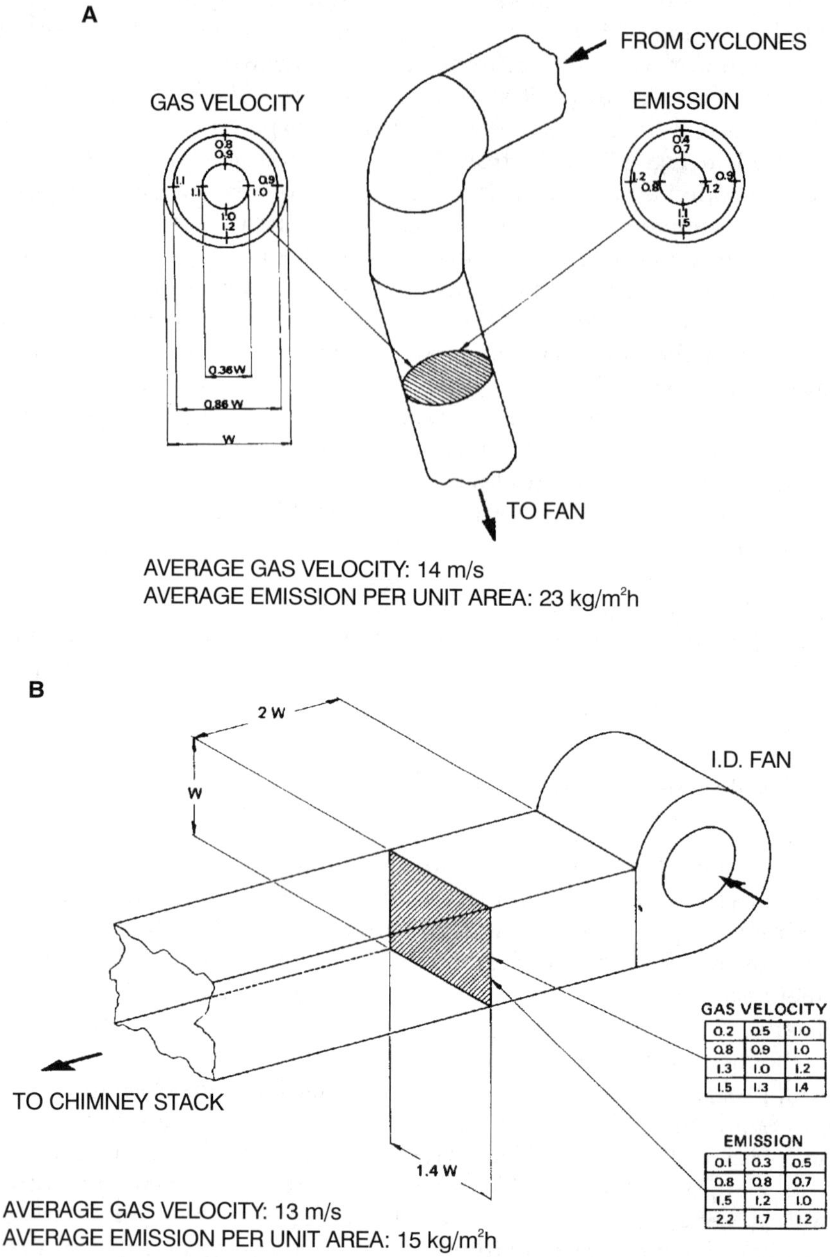

Fig. 3.2 Gas velocity and particle flux profiles from experimental measurements. (A) Downstream of a bend. (B) Downstream of a fan. (From Hawksley *et al.*, 1977; reproduced with permission of the Institute of Energy.)

	BS 3405	BS 6069/ISO 9096 Minimum	Ideal
Upstream of bend	*D*		2*D*
Downstream of bend	2*D*	*D*	7*D*
Fan	4*D*	4*D*	7*D*
Junction of two ducts	2*D*	*D*	7*D*
Fully open damper	*D*		
Partially closed damper	3*D*	3*D*	7*D*
Right-angle bend into chimney	2*D*		7*D*
Below chimney top	*D*		5*D*

US EPA Methods 1 (for velocity traverses) and 5 require the sampling site to be at least eight duct or stack diameters downstream and two diameters upstream from any flow disturbance. For a rectangular duct the equivalent diameter (also known as hydraulic diameter) is taken to be

$$D_e = 2LW/(L + W)$$

where L = length and W = width.

BS 6069/ISO 9096 specifies five hydraulic diameters downstream and two diameters upstream except when the sampling location is in the chimney when the requirement is increased to five diameters below the chimney top. It is recognized that the ideal requirement may not be met and the minimum requirements in such cases are shown tabulated above.

If it is impossible to find a suitable sampling location meeting the requirements of the relevant standard then it will be necessary to use a larger number of sampling points.

3.2.3 Number and relative location of sampling points

Having chosen the position in the flue where the sampling is to be undertaken, the number of sampling points across the flue can be decided. The principle is that each sampling point should be representative of an area which is a fixed fraction of the total flue area. BS 3405 recommends four or eight sampling points which are to be located as shown in Fig. 3.3. For rectangular flues, the distribution of sampling points is fairly obvious. Each point is representative of a smaller rectangular area 1/4 or 1/8th of the flue size. For small circular flues, each point represents a segment which is one-quarter of the circle. For larger flues, the inner four points are representative of segments of a circle having half of the overall area while the outer four points lie within an annular area representing the other half. The smaller number of points may be used if the flue area at that position is less than 2.5 m^2. However, if the velocity profile is very non-uniform (highest to lowest pitot reading greater than 4 : 1) the larger number of points is required.

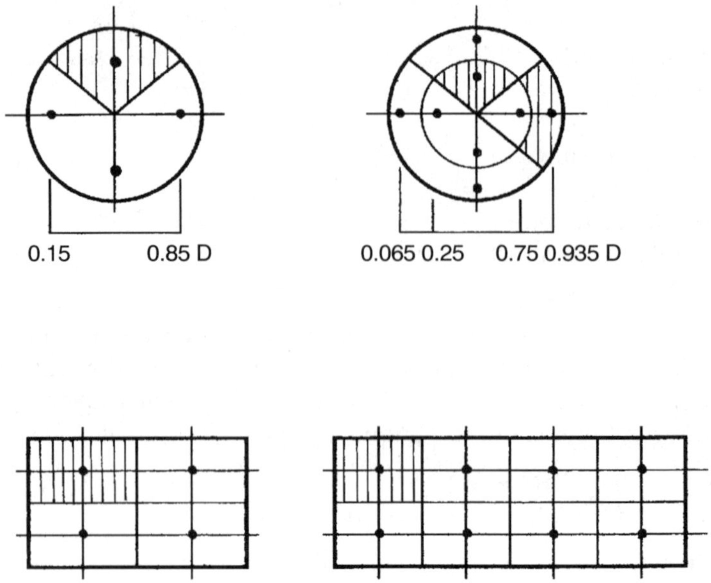

Fig. 3.3 Sampling points within circular and rectangular ducts according to BS 3405.

For greater accuracy – and especially if the requirements for sample location described above are not met – a larger number of sampling points should be used. For example, BS 893 would prescribe 12 sampling points for a 2 m × 1 m flue and 32 sampling points on eight different radii for a 5-m diameter circular flue. The recommended minimum number of sampling points for various standards are summarized in Table 3.2.

3.2.4 Isokinetic sampling

It is obviously important that the sample of gas drawn into the nozzle of the probe is representative of the gas in the flue at that point, otherwise the measured concentration will not reflect the actual concentration. Inertial effects can occur with large particles at the entrance to the nozzle, which can lead to serious error. These effects can be minimized providing that the velocity of the gas entering the nozzle is the same as the free stream velocity in the flue – this is called isokinetic sampling. The problem is illustrated in Fig. 3.4. If the velocity into the nozzle is too low (undersampling) there is a divergent gas flow. Some of the larger particles will not follow the streamlines as they pass round the outside of the nozzle but will carry on into the sampling system. The measured concentration is then too high. Conversely, if the sampling velocity is too high

Table 3.2 Recommended number of sampling points according to different standards

Circular flue			Rectangular flue		
Diameter (m)	No. of diameters	No. of sampling points	Hydraulic diameter (m)	Area (m^2)	No. of sampling points
BS3405					
[<1.8]*	2	4	[<1.6]	<2.5	4 (2 × 2)
[≥1.8]	2	8	[≥1.6]	≥2.5	8 (4 × 2)
BS6069/ISO9096					
<0.35	–	1	[<0.3]	<0.09	1
0.35–0.7	2	4	[0.3–0.61]	0.09–0.38	4 (2 × 2)
0.7–1.00	2	8	[0.61–1.2]	0.38–1.50	9 (3 × 3)
1.00–2.00	2	12	[>1.2]	>1.50	16 (4 × 4)
>2.00	2	16			
US EPA Methods 1, 1A, 5					
<0.3†	2	8	<0.3		9 (3 × 3)
0.3–0.61	2	8	0.3–0.61		9 (3 × 3)
>0.61	2	12	>0.61		12 (4 × 3)

* Diameters in brackets do not appear in the standards but are derived from the areas.
† Specific procedures for small ducts are given in Method 1A.

(over-sampling) the flow is converging but some of the particles will continue to the outside of the nozzle and not be collected. The measured concentration is then too low.

The procedure for adjustment to isokinetic sampling varies with the equipment used and further comments are given in sections 3.3 and 3.4 below.

There has been a considerable amount of theoretical study of errors arising from non-isokinetic sampling. The simplest approach is as follows (Hawksley *et al.*, 1977). Suppose one assumes that in the flue the true mass flux per unit area is m and the free stream velocity V. The sampling velocity and corresponding mass flux in the probe are V_s and m_s respectively. If all the particles followed the flow streamlines, then simple geometric arguments relate m, m_s and V, V_s: $m_s/m = V_s/V$. But if a fraction α of the particles originally carried with the gas does *not* follow this gas into the probe the relationship becomes

$$\frac{(m_s - m)}{m} = (1 - \alpha)\ \frac{(V_s - V)}{V}$$

In other words, the fractional error in the mass flux is proportional to the fractional difference between the sampling and free stream velocity.

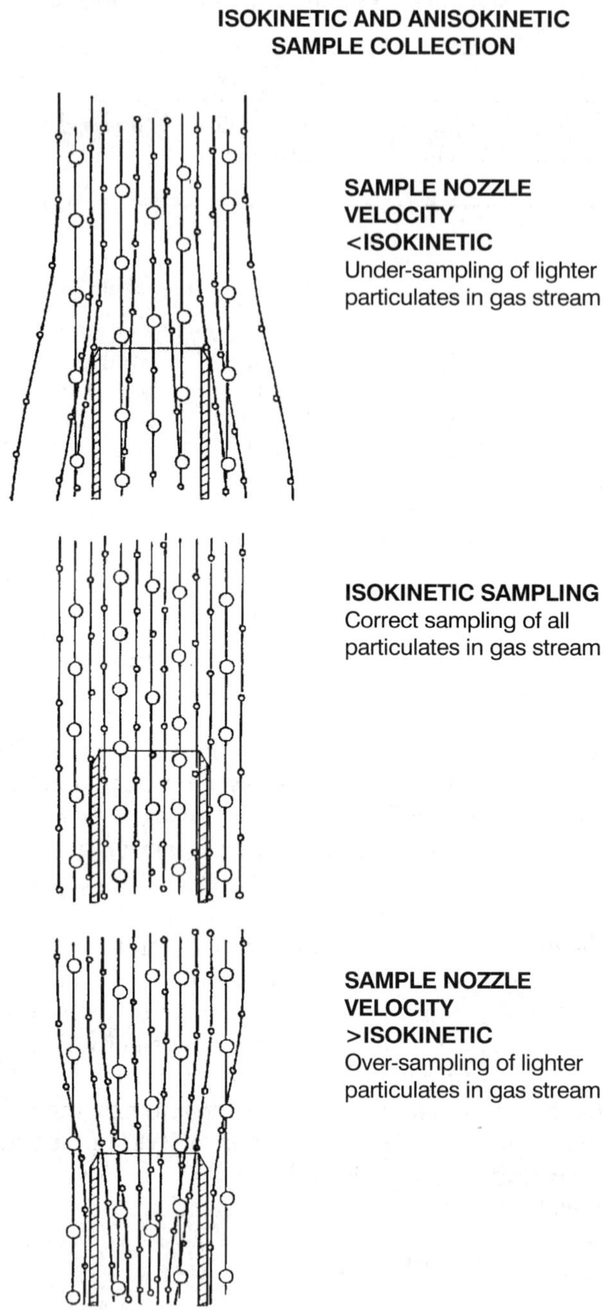

Fig. 3.4 The principle of isokinetic sampling.

The value of α of course depends on particle size, large grit particles having $\alpha = 1$ (when the mass flux will be correctly calculated but the estimated concentration could be severely wrong) while fume has $\alpha = 0$ and the concentration will be correctly calculated (whether you are sampling isokinetically or not). More detailed treatment of this problem relates to a parameter (VU/gD) similar to the Stokes number, where D is the nozzle diameter and U is the terminal velocity. In this context this is called the Barth No. (Ba). Figure 3.5 illustrates the errors that can arise due to non-isokinetic sampling in terms of Ba, and also the particle size for a typical density, gas velocity and nozzle size. As can be seen, for very large particles a factor of two error in the velocity gives rise to a factor of two error in the estimated particle concentration. The smaller the particles, the smaller the corresponding errors. Isokinetic sampling ceases to be necessary if one is dealing with particles which are below about 5 μm in diameter. This may be the case with some soots and fumes. However, care is necessary even if one believes the particle size is predominantly in this range because a small number of particles with large diameters can contribute a significant fraction of the total mass.

Even after adjustment to isokinetic sampling, errors may creep in if the nozzle itself is too blunt and causes disruption to the flow pattern. Geometrical requirements for the nozzles are given in BS 6069/ISO 9096 and EPA Method 5. A similar problem can arise if any other part of the sampling probe comes too close to the tip of the nozzle, whether upstream or downstream.

3.3 EQUIPMENT

3.3.1 Pitot tubes

The normal device for measuring the velocity in the duct is a pitot tube. The two most common types are illustrated in Fig. 3.6. The pitot static tube measures the difference Δp between the dynamic or velocity pressure and the static pressure in the duct. It has a single hole at the end of a tip facing into the flow and a series of holes around the tube at right angles to the flow. The pressure difference is given by

$$\Delta p = \tfrac{1}{2}\rho(V_{\mathrm{Duct}})^2$$

where ρ is the gas density and V_{Duct} is the duct gas velocity. The relationship rearranges to

$$V_{\mathrm{Duct}} = \sqrt{2\Delta p/\rho}$$

The velocity is related to the square root of the pressure difference which is measured either by an inclined-tube manometer or an electronic micromanometer. In practice the measured value is slightly different to

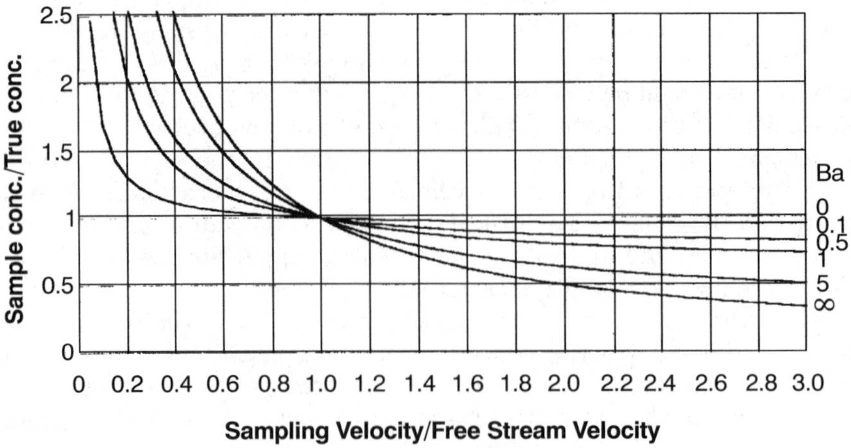

Fig. 3.5 Errors due to anisokinetic sampling as a function of the Barth No. (Ba). For free stream velocity 10 m/s, nozzle diameter 1 cm and particle density 2 g/cm³, the Barth Nos 0.1, 0.5, 1 and 5 correspond roughly to particle sizes 5, 10, 15 and 30 μm. (Based on Bohnet, 1978.)

the theoretical and this is allowed for by the addition of the pitot tube coefficient, C:

$$V_{\text{Duct}} = C\sqrt{2\Delta p/\rho}$$

The value of C depends on the precise design of the pitot tube – a hemispherical tip gives a slightly different value to an elipsoidal tip. The modified elipsoidal tip evolved by the UK National Physical Laboratory is recommended in British Standard BS 1042 part 2, 1983. This has C = 0.997, closer to unity than the spherical tip which gives C nearer to 0.99.

For work where the highest accuracy is not required, the density may be taken to be equal to the density of dry air at 1 atm corrected to the flue gas temperature T (Kelvin). The velocity is then proportional to the square root of the pressure differential and the temperature (see section 3.4.5 below). For more accurate work corrections should also be made for the static pressure not being 1 atm and the actual density or effective molecular weight M_e (relative molecular mass) not being identical to that of dry air. This requires some knowledge of the dry gas composition and the water vapour content:

$$M_e = \sum_i M_i V_i / 100$$

where i indicates the molecular species (N_2, O_2, Ar, CO_2, H_2O, etc.) and V_i the volume percent concentration. M_e(dry air) = 28.97. Then

$$\rho(\text{gas}) = \rho(\text{dry air})\, M_e(\text{gas})/M_e(\text{dry air})$$

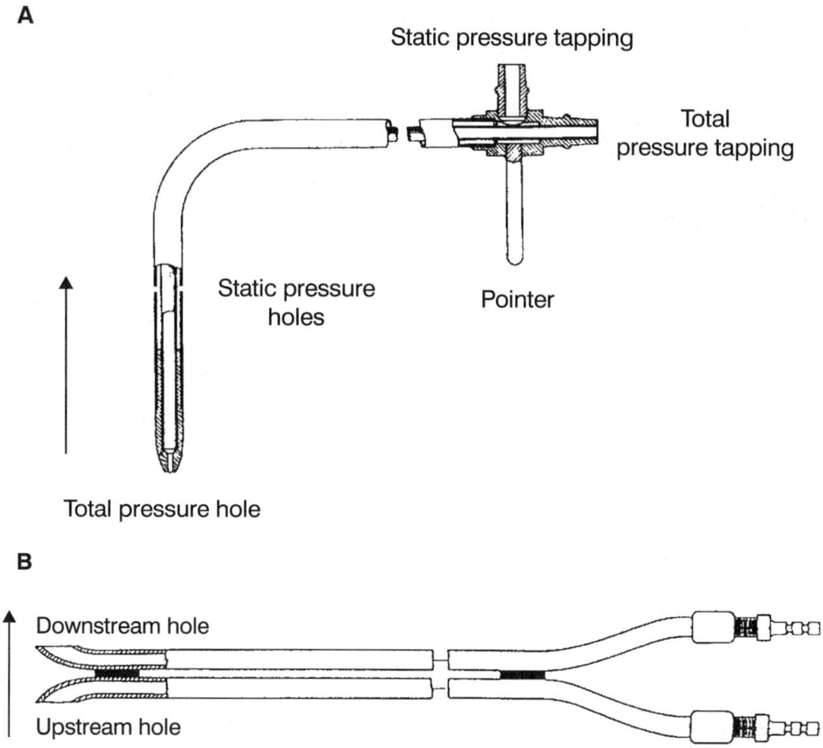

Fig. 3.6 Pitot tube designs. (A) Standard pitot static tube with ellipsoidal tip. (B) Stausschiebe (S-type) pitot tube.

The alternative design of pitot tube, called the S-type (Stausscheibe) or reverse type has twin nozzles facing one upstream and one downstream of the duct flow (Fig. 3.6B). In principle, this produces a higher pressure difference than the standard type; however, it needs calibration against the standard type. It has two particular advantages for stack gas monitoring. One is that the shape is more convenient, allowing the penetration of smaller-diameter holes in the flue wall because the nozzle-to-nozzle distance is less than the stem-to-tip distance in the standard type. It also has larger diameter holes which have less tendency to become blocked by accumulating particles or impinging droplets in the gas stream. It is therefore widely used and in the EPA Method 5 probe and some other commercial equipment is strapped to the particle sampling probe.

The pressure difference will only be accurately related to the velocity if the pitot tube is facing directly into the gas flow. As the angle between the flow direction and the axis of the nozzle (the yaw angle) increases response varies as shown in Fig. 3.7. There is a slight increase in differential pressure for angles in the range 0–15°, which is less than 0.5% for the standard type but can be several percent for the S-type. The pressure then decreases with increasing angle reaching zero at 90° for the S-type. The standard type gives a negative reading beyond about 60°.

One of the checks that must be carried out as part of the test procedure is to assess whether swirling or cyclonic flow is present. In other words, situations where the average gas flow direction is not parallel to the duct walls. These may occur downstream of certain duct configurations, of an

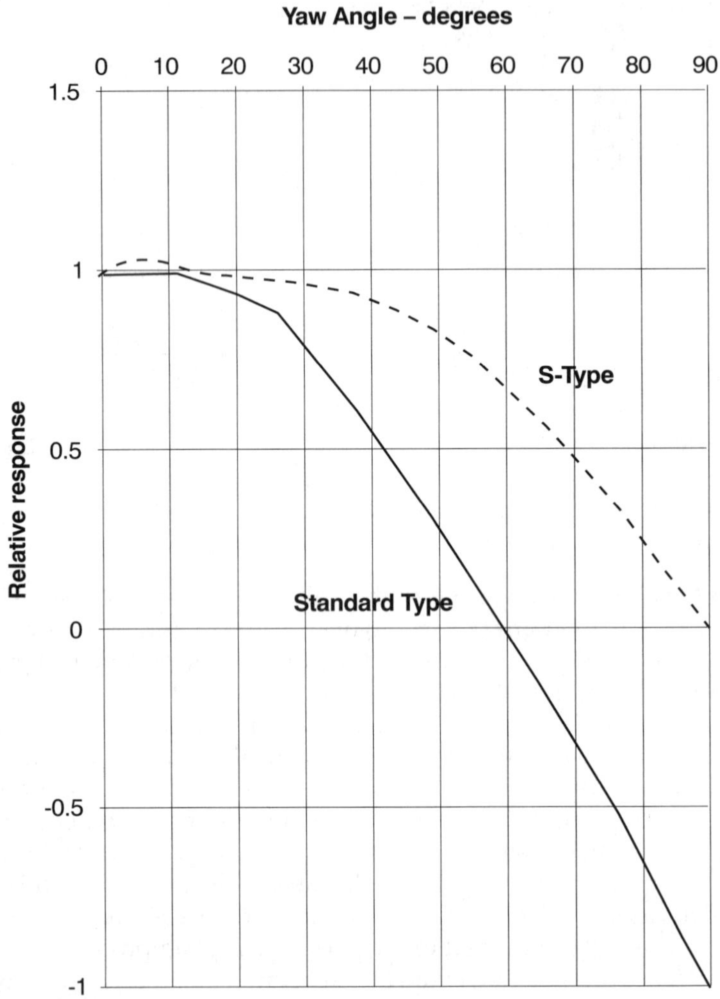

Fig. 3.7 Effect of yaw angle on the measured pressure difference with two types of pitot tube. These should be taken as illustrative, since the exact curves are velocity-dependent.

axial fan or of some demisting devices following a scrubber. The problem can be investigated by rotating the pitot tube and measuring the differential pressure at 5° intervals. In the absence of swirling flow the zero reading will be at 90° for an S-type tube. This will correspond to the maximum negative reading on a standard type tube. If the flow direction is more than 20° from the expected direction an alternative sampling location should be found.

Very low flow velocities cannot be measured by pitot tubes because the pressure difference becomes too low to measure accurately. BS 3405 suggests that pressures should be greater than 5 Pa (0.5 mm H_2O) or velocities greater than 3 m/s at 200°C. Velocities significantly below this may require the use of alternative techniques such as hot wire anemometry.

3.3.2 Sampling probes and sampling trains

The sampling train includes an entry nozzle, the probe tube and particle separation device followed by the gas flow control and metering equipment. The nozzle should be sharp-edged to avoid sampling errors and be mounted in such a way that it is upstream of any other component that might disrupt the flow. The particle separation device may be mounted in-stack or out of the stack. The in-stack arrangement (e.g. UK BCURA probe, US EPA Method 17 probe) has a number of advantages. The gas sample is filtered at the prevailing duct temperature and therefore there are no potential problems with condensation (except for very wet gases). Potential losses of particles in the probe tube are also reduced to a minimum. On the other hand, there may be materials problems with very hot locations. The BCURA probe can be operated up to about 400°C. Above this temperature the components of the probe can be rearranged to give external filtration (Fig. 3.8A,B). The latter arrangement must be used with care since there is no positive temperature control of the filter with this equipment. In contrast the (US EPA Method 5 probe (Fig. 3.9) uses external filtration with the filter mounted in a heated oven at 121°C (250°F). Provision is also made for heating the probe, if required, to avoid condensation.)

The particle separation device may involve single stage or two stages. With high particulate loadings filters can easily become overloaded and in these circumstances it is helpful to have an initial stage of particle removal in the form of a cyclone. Both the BCURA probe and the Method 5 probe (as an option) have this feature. The mass collected in the cyclone hopper as well as the mass on the filter must be measured. The filter can take many different forms. The BCURA probe uses a depth filter of packed glass or quartz wool in a stainless steel container. The disadvantage of this system is that it is not possible to undertake chemical analysis

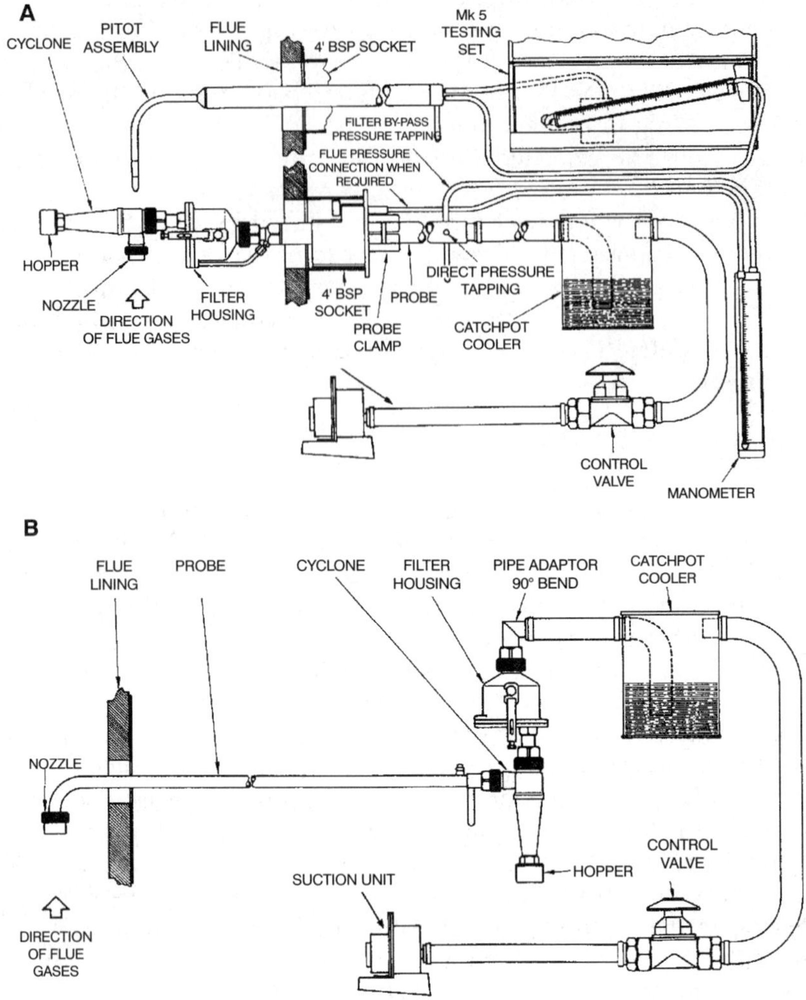

Fig. 3.8 The BCURA sampling system. (A) Normal operational arrangement. (B) High-temperature operational arrangement with the filter external to the flue. (Courtesy of Airflow Developments Ltd.)

of the collected particles, for example if metals content is required. For that purpose flat filter papers are most convenient and these are used in many probes. Another common design for in-stack samplers is to use filter thimbles which may be of paper, glass-fibre or, for very high temperature application, of porous ceramic. Figure 3.10 illustrates in-stack filter holders for use with flat filters and filter thimbles. The 'Mark 3A' sampler

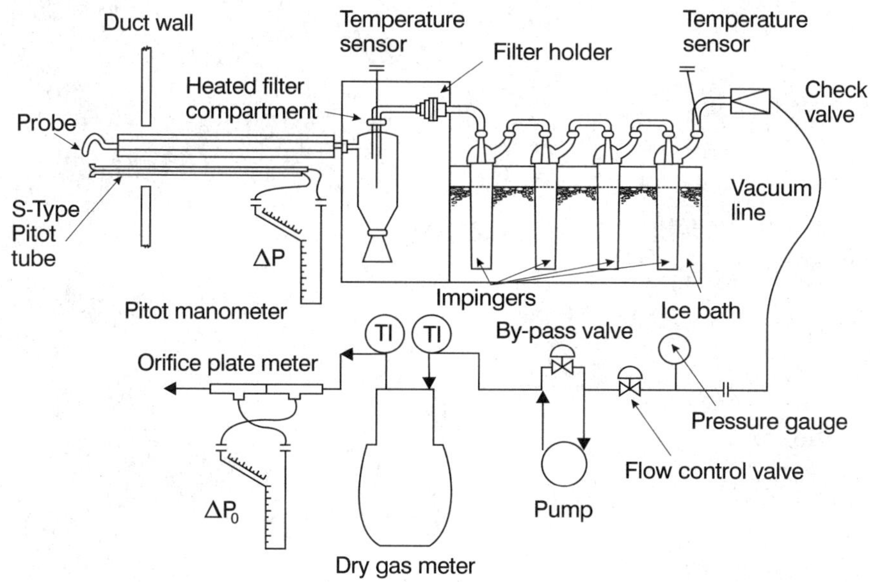

Fig. 3.9 US EPA Method 5 sampling train.

which has been widely used in the UK uses woven glass-fibre bags as the filter element.

Figure 3.8 shows the full BCURA (British Coal Utilization Research Association) sampling train excluding the temperature probe and pitot probe which are physically separate pieces of equipment. The probe is held rigidly in a clamp designed to be located in a 4-inch (10 cm) British Standard socket through the duct or chimney wall and enabling the probe to be slid in or out of the flue to the appropriate sampling point. Pressure tappings enable the pressure drop across the cyclone to be monitored to assist in adjustment to isokinetic conditions. A graph supplied by the manufacturer correlates the cyclone pressure drop with the measured pitot pressure difference at the same point. The flow rate is adjusted until the correct value of the cyclone pressure drop is obtained and the sampling velocity will then be isokinetic. Moisture is removed in a catchpot cooler and the remainder of the train simply involves a control valve and pump. In this system the sampled gas flow is not measured directly but estimated from the nozzle cross-sectional area and the gas flow velocity as measured in the gas stream.

Other sampling trains, such as the Method 5 train (Fig. 3.9), have both a dry gas meter (for total volume sampled) and a flow rate metering device such as an orifice plate plus manometer. Rather more careful moisture

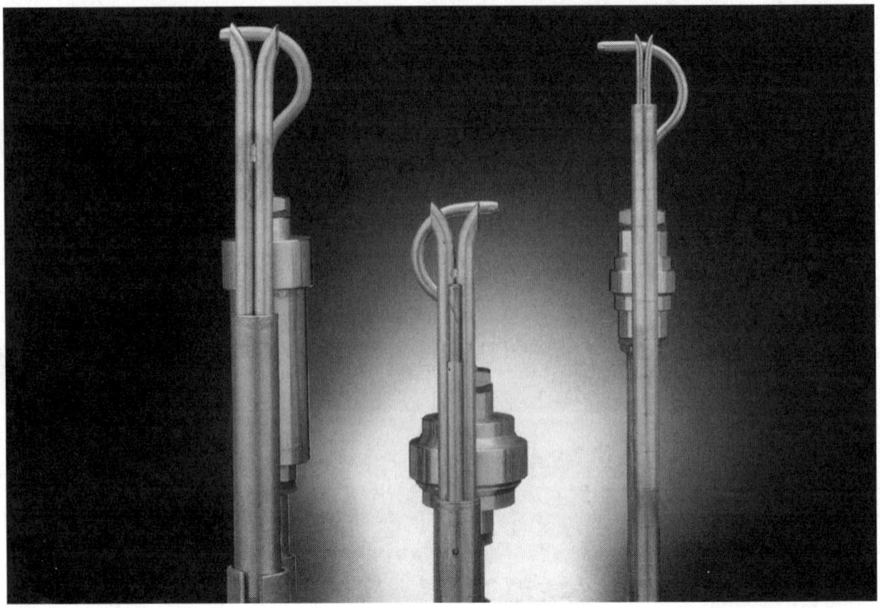

Fig. 3.10 Probe designs for in-stack sampling. Left: for filter thimbles; centre: for 47 mm flat filters; right: for 25 mm flat filters. (Zambelli s.r.l., Italy, courtesy of BIRAL Ltd.)

removal is required in such systems to avoid condensation in the gas meter and the gas pressure and temperature in the sampling line must be measured. The Method 5 system includes a series of impingers which allows the same equipment to be used for both gas and particulate analysis. When being used simply as a particulate sampling system the first two impingers are filled with water to remove most of the condensate, the third impinger is left empty and the fourth is packed with a solid desicant material.

By introducing electronic pressure measurement and control systems it becomes possible to automate the adjustment to isokinetic conditions. Various companies offer microprocessor-based automated systems which can log important parameters and undertake many of the necessary calculations relating to temperature, pressure, flow rates and duct velocity and provide a suitable print-out. After the sampling programme the filter weights are combined with the printed data to complete the calculation of emissions. The ultimate in automation is to use a motorized probe which can be programmed to move the nozzle to the required points in the duct, adjust to isokinetic conditions and take timed gas samples. However, although this level of sophistication is commercially available, it is as yet rather uncommon.

3.4 PRACTICAL ASPECTS OF SAMPLING

3.4.1 Pre-sampling preparation

Pre-sampling site visit

Pre-sampling site visits are an integral part of the preparation for sampling, during which much essential information may be obtained and used to prepare for a subsequent sampling exercise. The basic information which should be collected and recorded during such a visit is:

1. When and how the plant and process is being operated, e.g. what are normal hours of production, is the process a batch or continuous operation, is production affected by seasonal demand, etc.?
2. The nature and operation of the arrestment equipment serving the process, including indicating/recording instruments and maintenance schedules.
3. What records of plant, process and arrestment operating conditions are available and in what form? The operating conditions on most modern plants are now continuously recorded by computer systems, even if the plant is not fully computer-controlled. On some plants however, this may still not be the case and sampling staff should be aware of the need to record manually any necessary data where electronic down-load or print-out of information is not available.
4. Size and location of the emission duct to be sampled, particularly in relation to the final discharge to atmosphere. (The opportunity should be taken, where possible, to carry out exploratory pitot static tube traverses at the proposed, or where present, the existing sample plane, to check the flow profile of the duct gases.)
5. Access and sample platform facilities at the sample plane, including the provision of suitable power for use in operating the proposed sampling equipment.
6. Sample point facilities at the sample plane. (Where existing sample points are present, a physical check of the internal dimensions of each socket /fitment is essential. This should also incorporate 'freeing' of all sample point closures devices.)
7. The physical and chemical nature of the emission to be sampled, e.g. hot, wet, toxic, corrosive/acidic, under positive/negative pressure, etc..
8. What emission measurements are required to be made during the sampling exercise and what other samples of process streams are required?
9. Location of sample points and facilities for support sampling. (Fuel, product, waste streams/residues such as fly ash, etc.)
10. What analyses are required to be made on the samples collected during the sampling exercise?
11. Results of a comprehensive assessment of all safety aspects of the site and sample plane. (see also section 3.4.2).

Pre-site sampling – equipment and procedure preparation

The sampling equipment, in a clean condition, including the pitot-static tube and all pressure umbilicals, should be assembled and tested in the base laboratory to make sure that it is capable of use under the site conditions expected. An inspection of the equipment and a full leak test at this stage will often indicate if old, worn or damaged parts require replacement. All calibrations for the instruments to be used in the exercise must be current. It is particularly important that the total pressure inlet hole and nose of the pitot-static tube are free from damage. All the static pressure inlet holes of the pitot-static tube should be undamaged and free from blockages. The inlet edges of the sampling nozzles should be checked for wear or damage and defective items replaced. All sampling nozzle inlet diameters should be measured and accurately sized (using suitable Vernier callipers).

Filters and other physical collection devices such as hoppers should be prepared for the sampling exercise, by adequate conditioning followed by weighing to constant weight on suitable balances, including the inclusion of an adequate and representative number of blanks. It is good laboratory practice to standardize the physical process/processes used to condition filters and other physical collection devices, before weighing to constant weight. Due consideration however, must be given to the precise physical state of the particulate matter and any interfering material, expected to be collected during the sampling exercise, to ensure that no degradation of the samples occurs due to the reconditioning process. The weighed filters, hoppers and other collection devices should be transported to and from the site in suitable sealed bags/containers.

Where the sample train to be used incorporates 'wet chemical' arrestment the necessary collection equipment and solutions should likewise be made ready. (**N.B.** Where subsequent analysis of samples, from a sampling exercise, involves necessary checks on the efficiency of analytical recovery of collected material from the sample matrix, e.g. measurements of dioxins and furans, then 'spiking' of the collection media will be undertaken before conditioning and weighing.)

The keeping of adequate records is also a very important part of the monitoring/sampling operation and particularly so in relation to the analytical aspects of the work. The maintenance of an audit trail for all samples collected and analysed, arising from a sampling exercise is absolutely essential. This basically requires that each individual sample is given a unique identifier, to which all aspects of the sample collection and analysis, before and after the sampling exercise, can be related. The appropriate archiving of samples and records is also part of the maintenance of the audit trail.

3.4.2 Site preparation

This includes factors such as platforms, access, sampling points and safety. The provision of adequate and safe sampling platforms and access is an essential part of the preparations for sampling emissions to atmosphere. Unfortunately, it is also the one aspect of the preparation that is most often given the least attention. It cannot be stressed too strongly that without adequate and safe access the sampling work should not be attempted. A person's well-being or even life may depend on the quality of the sampling platform and access facilities provided.

The subject of the provision of sample platforms and access, sample points and facilities at the sample plane is considered in some detail in the HMIP publication, *Technical Guidance Note (Monitoring) M 1* (HMIP, 1990) and the reader is strongly recommended to consult this document before undertaking sampling work. The Health and Safety Commission are the responsible authority for all legislative matters relating to safety at industrial sites and premises and through their active arm, the Health and Safety Executive, apply the related regulations. The Health and Safety Executive are also responsible for producing a wide range of published material relating to all aspects of safety, among which are many useful publications relating to all aspects the work of sampling. The HSE address is given at the end of this chapter.

3.4.3 Sampling procedure

Arrival on site and transfer of equipment

After arrival at the site and the completion of the necessary introductions/safety induction, etc., the immediate task is the transfer of sampling equipment to the sample platform, followed by the assembly of the sample trains in preparation for the sampling exercise to commence. During this operation sampling staff will be observing the activity on the site/premises in order to ascertain that the existing site conditions do not present any new or unforeseen hazards in undertaking the work in hand.

Plant operation checks

After the equipment has been made ready for use, the next task is to carry out a check on the state and condition of the process and arrestment equipment. The rate/level of production that the plant and process is ideally required to operate at during the sampling exercise will, in all probability, have been determined during the preliminary discussions. It is imperative at this stage to ensure that the plant operating staff (through the correct lines of communication) are made fully aware of the requirements of the sampling exercise in relation to the agreed level of production, the maintenance of steady operating conditions throughout the sampling exercise, and the need to quickly make contact with sampling staff at the sample plane should any unforeseen circumstances arise which affect the operation of the process or arrestment plant.

Equally as important however, is that the process and arrestment plant can be confirmed as being at equilibrium, before the sampling exercise, for a sufficient period. This ensures that changes in the physical and chemical properties of the emissions, arising from that process during the period of the sampling exercise, are due only to the nature of the process – at equilibrium. Clearly, as is the case with all operating plant and processes, there will be many occasions when this ideal state of affairs cannot be achieved – usually for very good reasons – and sampling staff will then have call upon their experience, in order to decide the best course of action to take in the circumstances prevailing at that time. The essential point to bear in mind is the need to discuss fully and elaborate on such situations, in the resultant report of the sampling exercise. The most serious error of judgement that sampling staff can make, in any report of a sampling exercise, is to lose sight of the significance of the plant conditions, including any changes, during the actual sampling.

Isokinetic sampling procedure

No sampling staff who profess to be sufficiently experienced to undertake isokinetic sampling to any standard, be it British, European, American or otherwise, can hope to achieve credibility unless they have a thorough working knowledge and understanding of the standard to which they aspire to work. The procedural requirements in the UK for the sampling of emissions to atmosphere are laid down in BS 3405: 1983. For convenience, the main requirements for compliance with that standard are summarised in Appendix D. The principles on which it is based have been discussed earlier in the chapter and much valuable supplementary information is given by Hawksley *et al.* (1977). The supplier of the commonly used B.C.U.R.A. sampling equipment, Airflow Developments Ltd of High Wycombe, also issue a booklet describing the sampling procedure when using the equipment; their address is given at the end of this chapter.

Sampling is the most important part of the overall exercise of measurement of emissions to atmosphere, since, even the best analytical techniques available cannot correct for the effect of poor sampling practice or technology. Thus, the accuracy and precision by which the samples are collected, during the site sampling exercise, effectively determines the accuracy and precision of the end result of such an exercise. Modern analytical techniques provide resolution of samples to a significantly greater degree of precision and accuracy than those attainable for the collection of emission samples, even when employing the most sophisticated sampling methods and equipment available. A brief summary of the procedure is as follows:

1. Carry out preliminary 10-point pitot-static tube velocity pressure and temperature traverses, along each sample line, which include measurements at all the sample points to be used, in the sample plane.
2. Undertake isokinetic sampling at each sample point in the sample plane, based on the readings obtained in (1) above. (This may be cumulative or incremental, depending on circumstances).
3. Carry out repeat pitot-static tube velocity pressure and temperature measurements at all the sample points used in the sample plane.
4. Undertake repeat isokinetic sampling at each sample point used in the sample plane, based on the readings obtained in (3) above. (Using the same procedure as under (2) above.)
5. Carry out repeat pitot-static tube velocity pressure and temperature measurements at all the sample points used in the sample plane.

N.B. Measurements of the oxygen and moisture content of the duct gas, required to correct resultant emission concentrations to the requisite reference conditions, must be undertaken so as to be representative of the duct gas conditions throughout the sampling period.

3.4.4. Laboratory analysis

Good laboratory practice requires that incoming samples are received and subjected to all the required stages of the analytical process with due care and effectiveness. Documentation, as previously mentioned, is of prime importance in maintaining the audit trail for all samples.

Receipt of samples

The receipt of samples in the base laboratory involves initial checks on the integrity of the sealed containers followed by the matching of samples, presented for analysis, to the records of the samples associated with each sampling 'run', on the site sheets, making sure that all samples, including site blanks are accounted for. Laboratory staff at this stage will be aware of the analytical requirements of each sample and will initiate tracking of the samples, throughout the analytical procedures.

Conditioning and weighing

It is essential that all incoming samples intended for gravimetric analysis, including all site blanks, are re-conditioned before re-weighing to constant weight, employing the precise conditioning process/processes, used to prepare the filters and other collection devices, for the initial weighing process (refer also to Appendix D). It is important that all the weighing of filters and other sample collection devices, associated with a sampling exercise, should be undertaken on the one balance.

Further analysis

Following the completion of the required gravimetric analysis, any additional analysis of the sample collection should then be undertaken, to meet the sampling programme requirements. Where the additional analysis involves the alteration of the physical state, or destruction of the sample collection, e.g. preparation of solutions, extraction, etc. it is advisable, when practical, to archive part of the sample against the requirement of repeat analysis.

Reporting analytical results

When all the required analytical procedures have been completed, a thorough check of the results of all the analyses should be undertaken, to ensure wherever possible, that the integrity of the analytical processes can be assured. Repeat analysis of any doubtful results should be undertaken, whenever possible, before the results are passed forward for use in the report of the exercise.

It is useful at this stage to append comments on the analytical processes undertaken, for the benefit of the author of the report on the sampling exercise.

Archiving of samples

All the samples that have been collected from a sampling exercise should be archived into suitable storage, against future requirements and for a previously agreed period. Storage conditions should be such that there is negligible deterioration of samples for the period of retention. The proper archiving of samples is important, particularly so in those instances where the results of a sampling exercise may become the subject of legal action.

3.4.5 Calculations

The calculation of results from a sampling exercise needs to be accurate, complete and presented fully in the report. All the equations employed should be validated, before use, to ensure that all units and factors are consistent throughout and have been properly derived (metric units in Europe and the UK). 'Rounding-up' of decimal fractions should be used with care, after the effects on the results have been fully evaluated. All calculations carried out under site conditions should be checked for accuracy, before commencing on the final calculations, to be included in the report.

Velocity

The average velocity of duct gases, as measured at the sample points, is calculated from:

$$V_{Duct} = 0.0753 \sqrt{\Delta p}.\sqrt{T_{Duct}}$$

where V_{Duct} = velocity of duct gases, average, at sample points and at duct conditions (m/s); Δp = velocity pressure of duct gases, average (before sampling), at sample points (Pa, pascals); T_{Duct} = absolute temperature of duct gases, average, at sample points (K, Kelvin); and 0.0753 = Factor in consistent units.

Volume flow at duct conditions

The average volume flow of duct gases, as measured at the sample points, is calculated from:

$$Q_{Duct} = 60\, V_{Duct}\, A_{Duct}$$

where Q_{Duct} = volume flow of duct gases, average, at sample points and at duct conditions (m^3/ min); and A_{Duct} = cross-sectional area of duct, at the sample plane (m^2).

Volume flow at normal temperature and pressure (NTP)

The average volume flow of duct gases, as measured at the sample points and corrected to NTP (Normal Temperature and Pressure of 273K [nominally 0°C] and 101.325 kPa) is calculated from:

$$Q_{NTP} = \frac{Q_{Duct} P_{Duct} T_{Abs}}{P_{Abs} T_{Duct}}$$

where Q_{NTP} = volume flow of duct gases, average at sample points and at 273K and 101.325 kPa (m^3/min); P_{Duct} = static pressure of duct gases, average at sample points and at duct conditions (kPa); P_{Abs} = absolute pressure of 1 atmosphere (101.325 kPa); T_{Duct} = absolute temperature of duct gases, average at sample points and at duct conditions (K); and T_{Abs} = absolute temperature of 273K [nominally 0°C].

Volume flow at NTP and dry

Average volume flow of duct gases, as measured at the sample points and at NTP and dry, is calculated from:

$$Q_{NTP/Dry} = \frac{Q_{NTP}\,(100 - \%H_2O_{Duct})}{100}$$

where $Q_{NTP/Dry}$ = volume flow of gases, average at sample points, corrected to NTP and dry (m^3/ min); and % H_2O_{Duct} = average % moisture content of duct gases, measured at duct conditions.

Volume flow at reference conditions appropriate to the process

The average volume flow of duct gases, as measured at the sample points (at NTP, dry and corrected to the required reference % oxygen content for the process), is calculated from:

$$Q_{Ref} = Q_{NTP/Dry}\,\frac{[21 - \%O_{2Duct}]}{[21 - \%O_{2REF}]}$$

where Q_{REF} = volume flow of duct gases, average at sample points, corrected to NTP, dry and at reference conditions of % oxygen content, appropriate to the process (m^3/ min); % O_{2Duct} = average % oxygen content of duct gases, measured at duct conditions and, dry; and % O_{2REF} = % oxygen content of duct gases, at reference conditions appropriate to the process. (**N.B.** It is essential that the value for % O_{2Duct} used in this equation, is expressed on a dry basis.)

Mass emission

The average mass emission of pollutant, as measured at the sample points and arising from the release of the duct gases to atmosphere, is calculated from:

$$M = \frac{0.06wA_{Duct}}{t.a.n}$$

where, for cumulative sampling, M = mass emission of pollutant, average, as measured at the sample points and arising from the release of the duct gases to atmosphere (kg/h); w = total mass of pollutant collected, from all sampling points used during a sampling run (g); t = mean sample time at each sample point used, during a sample run (min); a = mean cross sectional area of sample nozzle inlet, used at each sample point (m^2); n = total number of sample points used for sampling; and 0.06 = factor in consistent units. (**N.B.** It is essential that '$a \times t$' remain constant at each sample point.)

Emission concentration

The average emission concentration of pollutant, present in the duct gases and as measured at the sample points, is calculated from:

$$C = \frac{16.67M}{Q_{REF}}$$

where C = emission concentration of pollutant, average, present in the duct gases and as measured at the sample points (g/m^3); and 16.67 = factor in consistent units. **N.B.** The value for the volume flow of duct gases, average, measured at the sample points and used in the above equation, may be on the basis of Q_{Duct};Q_{NTP}; $Q_{NTP/Dry}$ or Q_{REF}, depending on the definition required for the presentation of the results.

3.4.6 Report

The report of a sampling exercise, in either the draft or final form, should seek to present the overall results of that exercise in a clear and precise format. The layout should be structured in such a way as to take the reader through the whole report without the need to search back and forth for continuity. The contents should include all relevant information related to the exercise, thereby eliminating the need for the reader to request more details of relevant matters and reducing the potential for a wrong judgement being made, due to lack of the necessary detail. The report should not include statements that are based on supposition or unsupported opinion. The inclusion of published references in a report, to support the contents, should be such as to add to the quality of the work reported upon. Essential appended information should include copies of:

- all site sheets;
- all plant, process and arrestment operation records;
- all analytical data appertaining to the sample collection;
- all relevant records of instrument calibrations; and
- any photographs, relevant to the sampling exercise.

REFERENCES

Bohnet, M. (1978) in *Air Pollution Control, Part III* (ed. W. Strauss), John Wiley & Sons, New York, pp. 94–7.

British Standard BS 3405 (1983) Measurement of particulate emission, including grit and dust (simplified method), British Standards Institution.

British Standard BS 6069: Section 4.3: (1992) [ISO 9096: 1992] Characterisation of air quality. Part 4. Stationary source emissions. Section 4.3 Method for the manual gravimetric determination of concentration and mass flow rate of particulate material in gas-carrying ducts. British Standards Institution.

Hawksley, P.G.W., Badzioch, S. and Blackett, J.H. (1977) *Measurements of Solids in Flue Gases*, Institute of Energy, Second Edition, London.

Her Majesty's Inspectorate of Pollution (1993) Environmental Protection Act 1990, Technical Guidance Note (Monitoring) M1, Sampling Facility Requirements for the Monitoring of Particulates in Gaseous Releases to Atmosphere. HMSO, London.

Addresses

H S E Books, PO Box 1999, Sudbury, Suffolk CO10 6FS, UK. Tel: 0178 788 1165; Fax: 0178 731 3995. A price list of current H S E Publications is produced regularly.

Airflow Developments Ltd, Cressex Industrial Estate, Lancaster Road, High Wycombe, Bucks HP12 3QP, UK. Manufacturers and suppliers of the BCURA particulate emission sampling equipment and the CEGB Mark III A emission sampling equipment.

4

Particulate emissions – optical and other methods for continuous monitoring from a point source

Richard Horne

4.1 THE DEFINITION OF PARTICULATE EMISSIONS

All particles emitted from a stack or vent give rise to various nuisances, depending upon the size, quantity and chemistry of the particles. Plume visibility and haze are a major nuisance resulting mainly from the emission of finer particles, whereas larger particles give rise to local ground level deposits.

The emission of these particles to atmosphere is regulated according to their potential nuisance, and is expressed either in terms of the appearance of the plume (smoke density) or in terms of the mass of the particles per unit volume of exhaust gas (mg/m^3).

Current UK legislation suggests that for some processes the appearance of the plume should be controlled to a level expressed in terms of a Ringelmann number by comparing the greyness of the plume with a standard chart known as a 'Ringelmann' chart (Fig. 4.1), where 'clear' to 'black' is graded 0 to 5 in degrees of greyness (BS 2742: 1969 (1991)). This method has for years provided a standard for spot estimation of particulate emissions, which is quick and simple to effect.

However, although visual observations provide the simplest measurement they are highly subjective and prone to errors – particularly due to weather and daylight conditions.

Many other countries favour the use of the Bacharach system, where a controlled volume of the flue gas is drawn manually from the exhaust vent through a paper filter. The colour of the deposit collected on the filter paper is then compared with a Bacharach scale.

This is a quick and relatively simple technique and is ideal for assessing the amount of smoke produced by a combustion system, as carbon

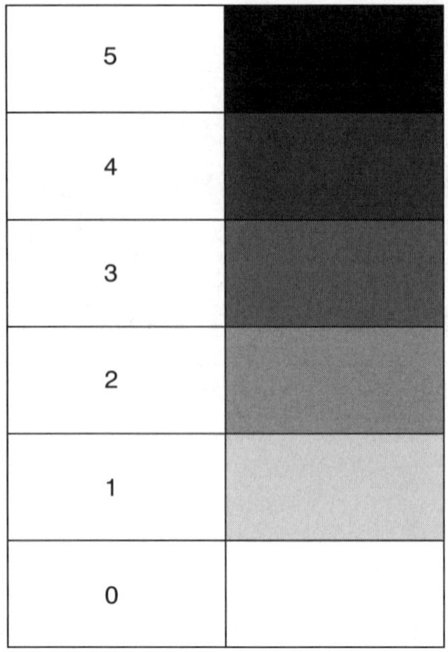

Fig. 4.1 Ringelmann chart.

particles have a uniform colour. However, it gives little direct indication of the final colour of the plume, and as with the Ringelmann chart, the result depends heavily upon the colour of the particles collected.

Where it is necessary to quantify particulate emissions in mg/m^3 a sample of the particulates should be collected from the exhaust duct isokinetically. Isokinetic sampling is accomplished by inserting an aspirating probe into the gas stream, which is operated in such a way that the sample velocity at the probe nozzle is equal to the gas velocity parallel to the nozzle. This ensures that the velocity profile of the gas stream is unmodified by the presence of the probe and that the collected sample is representative (Fig. 4.2).

The sample is extracted continuously over a fixed period of time, typically between 10 and 30 minutes, and the particulate content collected on a filter which is weighed before and after the sampling period. From a knowledge of the sampling flow rate, a value of solids emission in mg/m^3 may be obtained. Further details of extractive sampling are given in Chapter 3.

It is important, when describing the density of particulate (and gaseous) emissions, that all measurements are expressed at the same reference conditions. These are typically defined as follows:

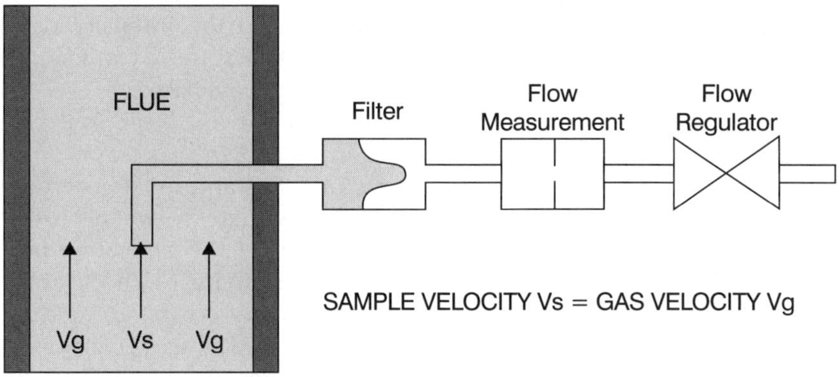

Fig. 4.2 Isokinetic sampling.

Temperature	273K
Pressure	101.3 kPa
Oxygen level	3%,6% or 11% dry (depending on the process)

The measurements are usually also expressed on a 'dry gas' basis (i.e. with the effects of water vapour removed). Measurements defined at reference conditions are usually expressed in mg/Nm^3 or $mg{\cdot}Nm^{-3}$ (i.e. milligrams per 'normal' cubic metre). Normalization allows emissions from different processes to be compared even if they are at different temperatures and with different amounts of dilution.

Although visual observation of the plume and manual collection of the particles unarguably provide the most accurate means of analysis, they are, by definition, historical and give little information about the dynamics of particulate emissions. They cannot therefore be considered for continuous monitoring to show compliance with the demanding legislation which is already in force in many countries.

4.2 DESIGN OF OPTICAL DEVICES FOR CONTINUOUS MEASUREMENT

4.2.1 Conventional transmissometers

International legislation suggests the use of an optical measuring device for the continuous assessment of particulate emissions. Such devices are known as opacity monitors, opacimeters, transmissometers, smoke density meters, etc.

Simple, single-pass transmissometers provide the most obvious form of continuous monitor. They consist of a light source mounted on one side of the duct/chimney, projecting visible light across the duct towards a receiver on the other side of the duct. Particulates in the beam at that point absorb and scatter light causing changes in the intensity of the received light. This arrangement is shown in Fig. 4.3. If we can keep the optical surfaces clean, this is the simplest technique available.

High-efficiency air curtains must be used to prevent gas-borne particles reaching the optical surfaces. Any residual contamination on optical surfaces will increase the apparent particulate level and must be avoided because simple transmissometers cannot differentiate between particles held in the gas stream and particles deposited on the optical surfaces. More intelligent systems can use periods of inactivity to check and, if necessary, automatically reset zero to eliminate any errors. On continuous processes with continuous particulate emissions however, errors can only be established by removing simple systems from the duct and checking the zero point on a rig which simulates the clear duct.

In an attempt to resolve the problems of the simple single-pass transmissometer, double-pass (folded beam) systems were produced, which have the light source and the receiver mounted on the same side of the duct. The light is focused across the duct onto an auto-collimating reflector which reflects the beam back, through the lens assembly again and onto the receiving detector element (Fig. 4.4).

Doubling the path of a transmissometer effectively doubles the sensitivity and, by periodically isolating the instrument from the effects of the

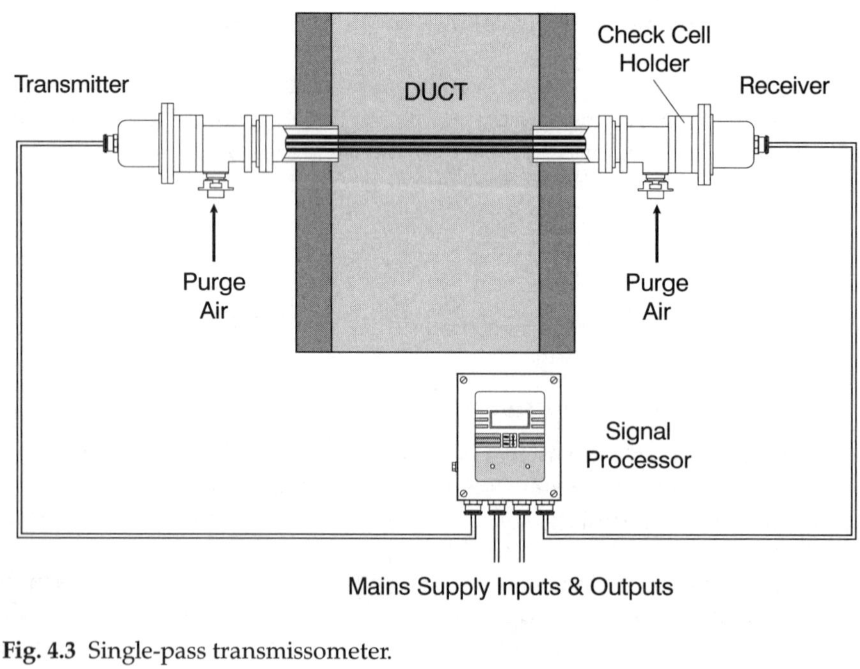

Fig. 4.3 Single-pass transmissometer.

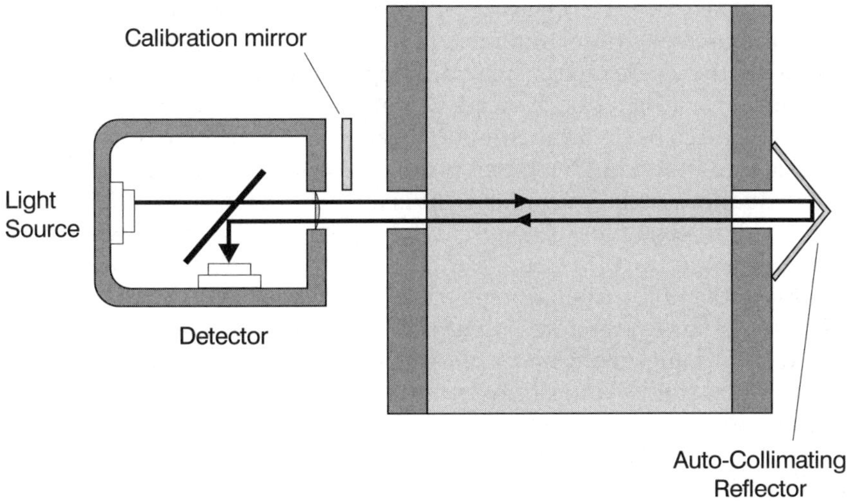

Fig. 4.4 Double-pass transmissometer.

process gas by inserting a mirror directly in front of it, a zero condition can be simulated and a measurement made of any contamination on the window of the transceiver.

Unfortunately, this procedure cannot account for dirt on the retro-reflector and an assumption has to be made that contamination of the retro-reflector will be the same as that of the lens. This can be a source of zero errors if turbulent flow contaminates only one side of the duct or if one of the air curtains is not functioning correctly.

With both single- and double-pass transmissometers, there is no automatic means of checking the optical alignment. If the alignment shifts significantly, only part of the transmitted energy may be collected and the reduction in received energy will be considered by the instrument as an increase in particulate density. During calibration, double-pass transmissometers not only eliminate the effect of the particles within the duct but also eliminate the effect of the reflector. They cannot therefore detect this potential error.

Well-maintained double-pass systems can provide significant benefits over single-pass systems, but they still fall short of resolving all of the potential calibration problems of transmissometers.

Double-beam transmissometers (Fig. 4.5) use a combined light source and detector unit mounted on both sides of the duct to be measured. The units are identical and both of them transmit and receive light alternately many times every second.

Each of the double-beam transceivers constantly monitors received light intensity, which is either a measure of opacity (when the opposite transmitter is ON) or a measure of the intensity of its own light source (when the opposite transmitter is OFF).

As with the double-pass system, integral mirrors can periodically be positioned in front of each head to remove the effects of the particles within the duct, but with the double-beam system the contamination can be measured individually at each head and compensated for.

This dual transceiver arrangement is also able to detect misalignment. Each head makes a measurement at the same point but in different directions across the duct. Measurements made in the two directions are different if intolerable misalignment occurs.

This technique therefore appears to resolve the potential problems caused by misalignment and optical contamination that are inherent in conventional single- and double-pass devices. It eliminates all the major potential sources of error and provides what is generally considered to be close to the practical limit of accuracy for optical transmissometers.

Although conventional transmissometers provide the accepted and most common technique for continuous particulate monitoring, they will not provide the accuracy required if they respond to anything other than the particulates in the duct. A transmissometer should therefore have a stable, controlled source which should be uniform to make it tolerant to changes in alignment. It should also be insensitive to surrounding light conditions – especially when measuring in short, wide chimneys where the total received energy is often the sum of both the transmitted energy

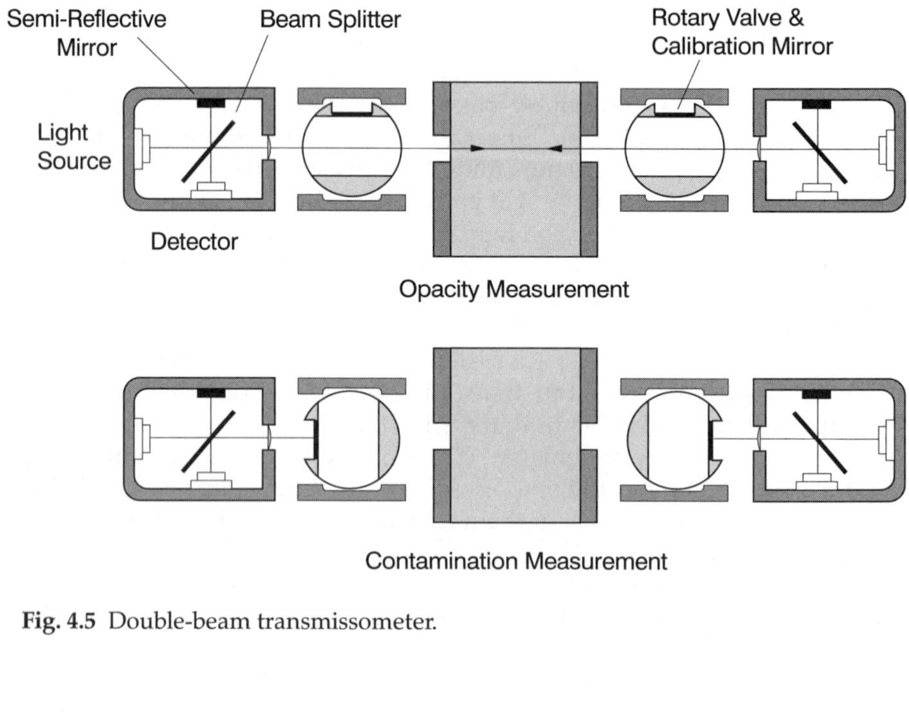

Fig. 4.5 Double-beam transmissometer.

and daylight. To operate under such conditions instruments must transmit coded energy (usually pulsed at a specific frequency).

The implementation of all these design criteria into a system can make it relatively expensive, especially for smaller processes. A much less expensive technique is to allow these parameters to drift and only measure the short-term fluctuations produced by changes in the dust density.

The background drift produced by optical contamination or shift in alignment is usually long term; therefore, for low-level particulate emissions from processes with fabric filters, this simpler technique provides an attractive filter failure device. However, because these systems only assess the dust burden against short-term fluctuations in transmission, there can be pitfalls.

Combustion processes which are capable of generating smoke can produce a real high level background of suspended particles for a relatively long period. Such particles will be discounted by these deviation transmissometers even though they undoubtedly contribute to the total particulate burden.

4.2.2 Opacity fluctuations or 'flicker' effect

Some transmissometer systems rely totally upon a 'flicker' component of the optical transmission (Fig. 4.6), which is said to correlate with the dust density. This is sometimes called scintillation. The author has witnessed such systems operating on stable processes with a stable particulate definition and their output appears to reflect the process conditions. Unfortunately, technical papers in support of this technique are still rare. A recent paper by Clarke (1996) has described the theory of the instrument and the factors which influence the output signal. However, the ability of such instruments to measure variable particulate burdens consistently at variable velocities and in different types of plant has yet to be established.

4.2.3 Back scatter devices

The change in optical transmission for a fixed particulate density is a function of the optical path length. Dirty gases in large ducts attenuate more energy than they would in small ducts. For super low level detection, especially in very small ducts, transmissometer technology therefore tends to be less successful and it is necessary to consider other techniques.

Back scatter devices can provide this sensitivity as they do not look for changes in optical transmission but look for light reflected back from suspended particles illuminated by a bright source (Fig. 4.7).

The phenomenon has been observed by all who visit the cinema. The dust particles in the atmosphere of an apparently 'clean' cinema are

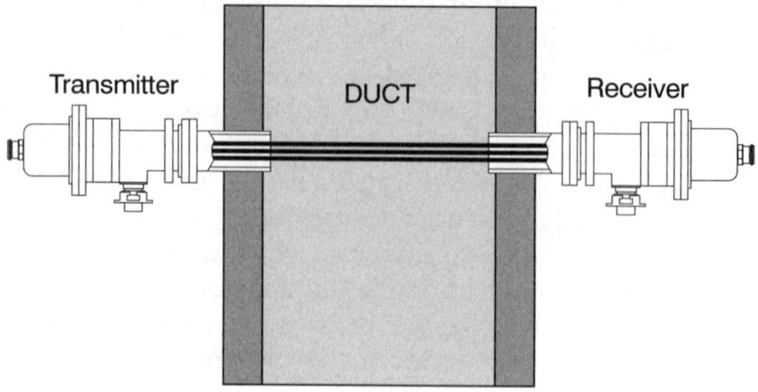

Fig. 4.6 Flicker (or scintillation) effect monitor.

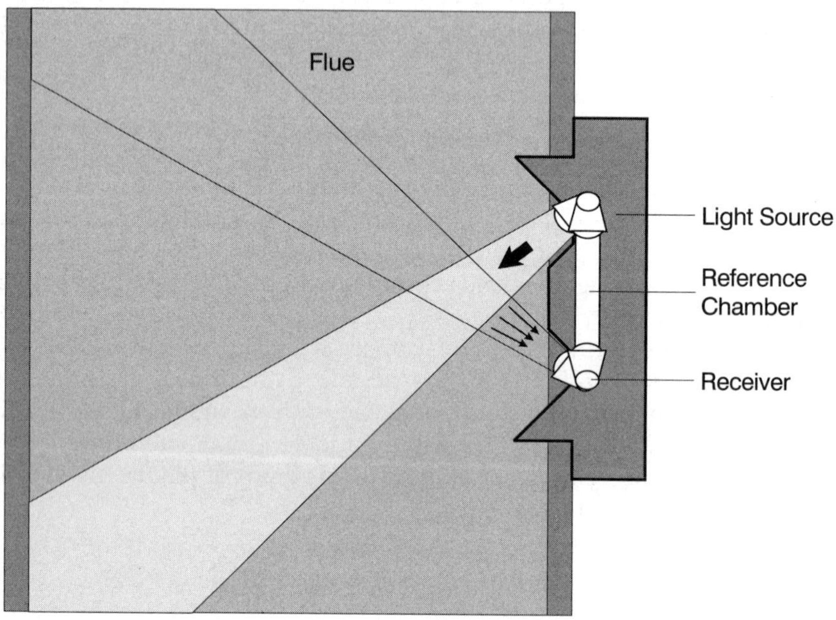

Fig. 4.7 Back scatter device.

suddenly highly visible when the cinema lights go down and the powerful projector beam illuminates them against the now dark background.

Back scatter devices are indeed very sensitive – capable of measuring down to micrograms per m^3. They are however, not generally suitable for use on large ducts as they only measure a local plane close to the duct wall and maximum ranges of only 0–20 mg/m^3 are often quoted.

4.2.4 C.E.R.L. dust monitor

Another optical technique used for assessing particulate emissions from process plant is used by the C.E.R.L. (Central Electricity Research Laboratory) dust monitor (Fig. 4.8) (Lucas *et al.*, 1972).

This system considers particles in terms of their ability to be deposited and, as such, is more suited to the analysis of large particles. As transmissometers are less sensitive to large particles than to smaller ones, this technique has been popular for monitoring emissions from processes such as oil-fired power stations where particles greater than 200 μm are often encountered.

4.3 OTHER TECHNIQUES FOR CONTINUOUS MEASUREMENT

4.3.1 Triboelectric probe

Another device for continuously assessing the particulate emissions from process plant uses a metal probe which is positioned in the gas

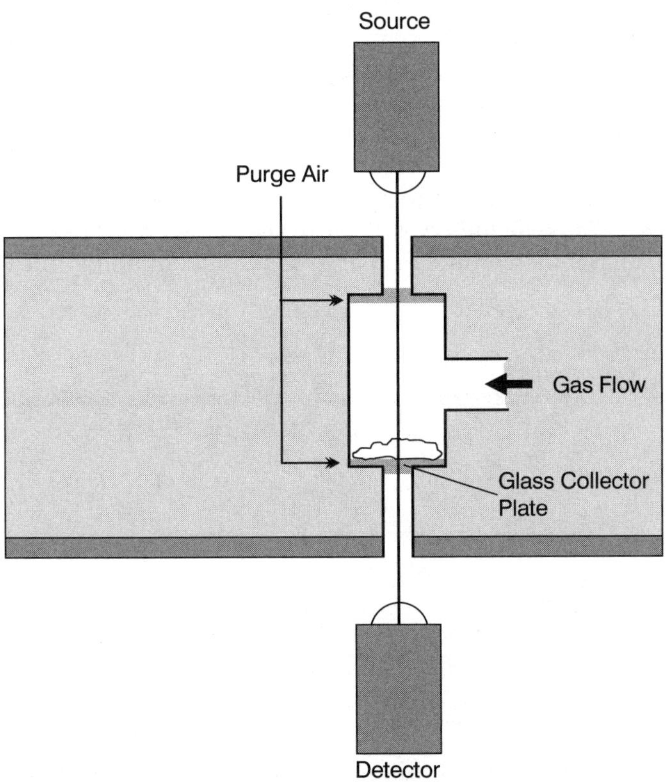

Fig. 4.8 C.E.R.L. monitor.

stream (Fig. 4.9). As particles impinge on the insulated probe, an electrostatic charge is either donated to the probe by the particle, or donated to the particle by the probe. Such devices are variously described as:

- triboelectric probes
- charge transfer probes
- particle impingement probes.

Their concept is initially attractive because they require access to only one side of the duct and are not significantly affected by a build-up of particles. They are sensitive enough for low-level detection and often used as 'breakthrough' detectors for the measurement of particles on stable, fixed flow rate processes, especially after fabric filters.

For quantitative analysis, they must be positioned in ducts with consistent flow characteristics to prevent particles from having more than one opportunity to make contact with the probe (Fig. 4.9). A change from laminar to turbulent flow can produce a dramatic increase in output for the same particulate density.

These probes provide a basic measurement which is proportional to mass per unit of time. If they are to be used on processes with variable flow rates, it is necessary to measure continuously and correct for pollutant gas velocity in order to express particulate density in mg/m^3 as required by legislation.

Finally, as charge transfer only occurs through contact with the probe, consideration must be given to applications with very small particles and at relatively low velocities. Small particles at low velocities can operate almost as a gaseous medium and are often reluctant to be thrown out of

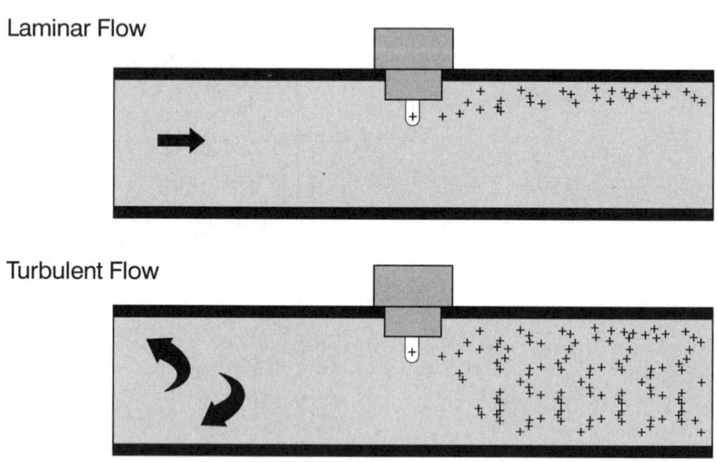

Fig. 4.9 Triboelectric probe.

suspension to impact on the probe. If a boundary layer builds up around the probe, this can deflect such particles and cause them not to donate/receive a charge.

The techniques described here and the many other derivatives not detailed are all inferential techniques. That is, they measure some parameter which, within certain limits, can be used to assess the particulate density. They are all however, prone to various errors if the physical and chemical nature of the particles vary significantly.

4.3.2 Beta gauge sampler

One technique which is claimed to be totally independent of the physical and chemical properties of the particles uses beta-radiation (Fig. 4.10). It is not an *in situ* device but collects a sample of the flue gases isokinetically via a probe in the duct. A continuous roll of filter medium is used to trap the particles over a controlled period of time and the collected particles are measured using an integral beta-radiation source and gauge.

Particulate matter attenuates beta-radiation according only to its mass (and not its chemical composition). Therefore, if the mass of the relevant section of filter is measured before and after it has been used, the mass of the collected particulates can be established. With a knowledge of the rate of gas collection this mass can be expressed in mg/m^3.

Where particle size and composition variations are great, beta gauges can apparently provide the solution, yet, under these conditions, distribution of the particles throughout the duct is at its least predictable and emission densities often vary by as much as ±200% across the duct.

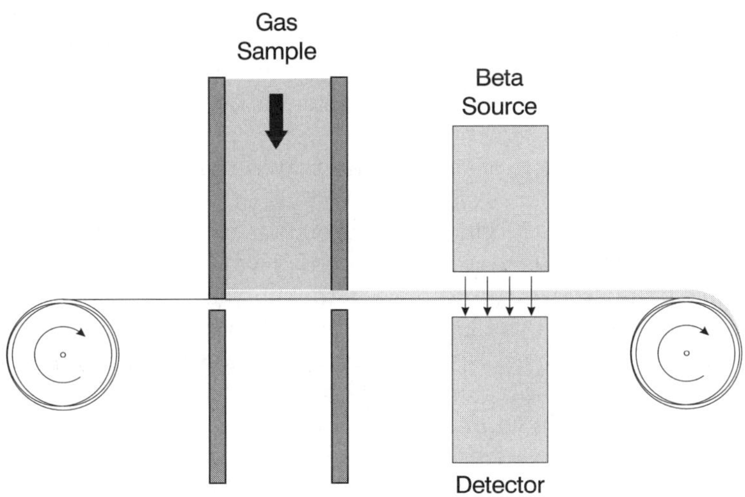

Fig. 4.10 Beta gauge.

Using a single sample probe only allows a sample to be collected from one point in the duct; therefore, selecting the best (most representative) point to sample is often impossible – especially with varying load/flow conditions.

It is also extremely difficult with any extractive system to guarantee that all particles actually reach the filter medium. Deposition of particles within the probe and sample train result in both errors in the measurement and a proportionately higher degree of maintenance compared with non-contact techniques.

It is interesting to note that the deposition of particles in sample lines undervalues beta gauge measurements, whereas the deposition of particles on the optical surfaces of transmissometers overvalues the measurement.

In conclusion, no single technique stands out as being perfect for all applications. Because optical transmissometers consider the entire duct, can be used to assess the appearance of the plume, are unaffected by turbulence and velocity, and are relatively simple to apply and maintain, they are still the most popular technique. They are also the preferred technique for the majority of international legislation and the standard against which all other techniques must be compared.

4.4 INTERACTION OF LIGHT WITH PARTICLES

4.4.1 Background

If measurements made by optical transmissometers are to be consistent, it is necessary to ensure that they are made always at the same wavelengths. It is convenient to take the response of the human eye as the standard, since this is the detection system used for assessing the appearance of the plume.

The sensitive region of the eye lies between 400 and 700 nanometres (nm). Such a response is said to be 'photopic' and in order to maintain comparability between optical measurements it is essential that any optical system used has its response within this waveband.

Visible light also has the advantage that it is not absorbed by major flue gas constituents such as CO_2 and water vapour. It is however, significantly affected by visible gases such as NO_2, and for measuring particulates, the 'red' part of the spectrum should be considered in order to minimize this effect. (This is in contradiction of the USA EPA requirement, which insists upon measurement in the 'green' part of the spectrum – where sensitivity to NO_2 is at its maximum.) Visible light is also attenuated by water droplets; therefore, condensates or aerosols carried over from wet scrubbers must be avoided at all costs.

The attenuation of light by particulate matter is further influenced by the wavelength of the light and the size of the particles (Fig. 4.11) because the scattering ability of the particles reduces as their size decreases below the wavelength of the light used. Thus, long-wavelength or infra-red light is unable to detect small, submicron particles as efficiently as visible light.

At this point it must also be recognized that optical methods can only be applied if the particulate size is small enough for the mass of the solids to be evenly spread. Provided that the particle size criterion is met, there are certain very specific relationships between optical measurements and the particulate concentration.

4.4.2 Relationships between optical measurements, plume appearance and mass emission

Before we consider the relationships between optical measurements and suspended particulates in a duct, we must first consider the following common terms used in optical measurement:

Transmittance: this is the ratio between transmitted energy and received energy and is calculated thus:

$$\text{Transmittance } T = \frac{I}{Io} \tag{4.1}$$

where Io = intensity of light transmitted; and
I = intensity of light received.

Opacity: this is the reduction in transmitted energy expressed as a percentage:

$$\text{Opacity} = 100\,(1 - T) \tag{4.2}$$

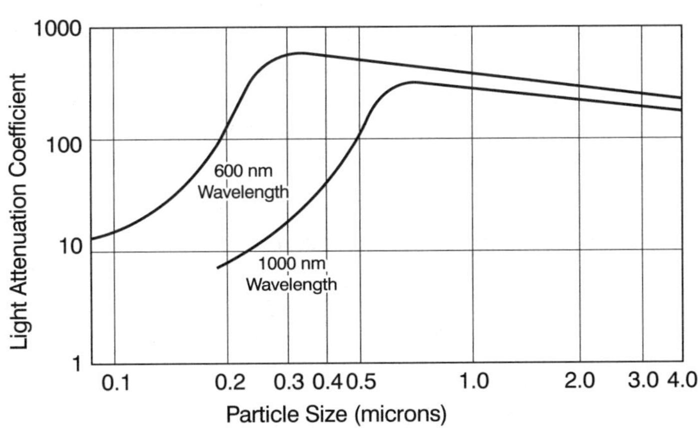

Fig. 4.11 Attenuation of light.

Extinction: this is a function of transmission which (as explained later) is proportional to particulate concentration:

Extinction $= \log_e (1/T)$ (4.3)

The output from simple transmissometers is usually expressed in terms of % opacity (or % smoke density). Transmissometers used for the assessment of particulate mass emission usually provide optional outputs in both opacity and extinction. To add to the confusion, note that extinction is also often referred to as 'optical density' and sometimes calculated as $\log_{10} (1/T)$.

In order to assess the appearance of the plume continuously and under all weather conditions, the relationship between Ringelmann number and opacity is described in BS 2742:1969 Addendum 1 (1972), which assumes a 'standard' emission against a 'standard' sky. The relationship is defined as follows:

Ringelmann 1 = 52% opacity
Ringelmann 2 = 64% opacity
Ringelmann 3 = 76% opacity
Ringelmann 4 = 88% opacity

Unfortunately, the British Standard fails to define a range of stack diameters for which this relationship applies and the author wishes to comment that on small-diameter stacks, 52% opacity in practice often looks like very dark smoke.

It would appear, therefore, that if Ringelmann shade is to continue being used for continuous assessment of emission performance, further work is required in order to define the relationship fully, unless each point of emission is to be characterised individually. The US system of defining acceptable emissions in terms of % opacity is much more sensible.

Finally, stack opacity (and Ringelmann) is an assessment of the appearance of the plume at the stack outlet. It is often more practical, however, to measure at some other point for ease of access. If the duct diameter at the measurement point differs from that at the stack exit, the following corrections must be applied:

$$O_x = 100\,(1 - (1 - O_m/100)^{Lx/Lm})\ \%$$

where O_x = effective opacity at stack exit;
O_m = measured opacity;
L_x = stack exit diameter; and
L_m = path length at measurement.

Usually there is little change in temperature between the measured and exit points but where it is appreciable, the stack exit opacity will be modified. This is due to the gases reducing in volume while the particulate volume remains relatively constant.

This correction is applied by expressing the effect in terms of the measurement path length as follows:

$$L_m' = L_m \frac{(273 + T_x)}{(273 + T_m)}$$

where L_m' = temperature corrected effective measurement path;
L_m = actual measurement path length;
T_m = temp (°C) at measurement; and
T_x = temp (°C) at stack exit.

Figure 4.12 shows the effect of some variations in stack exit versus measured path ratios.

If we wish to use an optical device for assessing mass emissions, we must consider the manner in which light is attenuated by fine particulate matter. The relationship is defined mathematically by the Beer–Lambert law:

$$\text{Transmittance } T = \exp(-k \,.\, n \,.\, a \,.\, l) \tag{4.4}$$

where k is the extinction efficiency of the particles;
n is the number of particles per unit volume;
a is mean projected area of the particles; and
l is the path length through the gas.

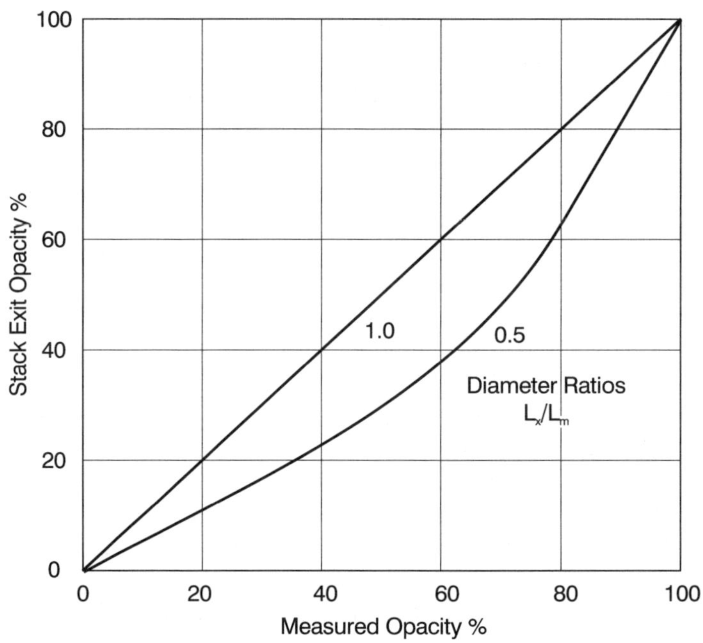

Fig. 4.12 Measured versus stack exit opacity.

The extinction efficiency, k, determines precisely how much light is attenuated by the particles and depends totally on the physical and chemical nature of the particles. For continuous mass emission, a measurement of extinction is convenient because:

$$\text{Extinction} = \log_e (1/T) = k \, . \, n \, . \, a \, . \, l$$

This parameter is useful in that, assuming the nature and size distribution of the particles are constant, for a specific path length, extinction is directly proportional to the particulate concentration n, and hence to the particulate mass emission in mg/m^3.

It is interesting to note that, for a fixed particulate density, extinction increases as the measurement path length is increased. This makes transmissometers ideally suited for the measurement of suspended particulates in large ducts.

4.4.3 Calibration

The unknown parameter in the Beer–Lambert relationship is the extinction efficiency, k, which can only be established empirically. In practice, this and the other constants, a and l, are established by calibrating each instrument against specific measurements of solids emissions, by the isokinetic sampling technique described earlier. The resulting calibration value would hold only for that particular flue and would need to be rechecked at regular intervals by further sampling comparisons.

If, for a particular measurement point in a particular flue/duct/chimney, we collect an isokinetic sample over a defined period, and over the same period a continuous monitor measures and calculates the equivalent average extinction, the relationship between these measurements may be used to produce a 'Dust Factor' where:

$$\text{'Dust Factor'} = \frac{\text{measured fine dust density}}{\text{measured extinction}}$$

This factor can then be used to calibrate an instrument directly in terms of dust density (mg/m^3).

Such calibration is required for all continuous particulate monitors – whatever their operating principle. For instruments with a linear scale shape, and on processes where the particulates are stable in size and chemical composition, a single isokinetic test can provide enough data to allow measurement across a limited range of emitted particulate concentrations.

For instruments with non-linear scale shapes, or where the particle size and composition vary significantly, it will be necessary to perform a series of tests and, if necessary, to characterize the measurement according to each specific application.

One aspect of monitoring particulates which is often overlooked is the need on some processes to correct measurements to standard conditions, especially for emissions from combustion processes.

Just as any collected sample will be expressed at standard (normalized) conditions, the output from any continuous particulate monitor might also require correcting for variations in diluting factors such as temperature, excess air (O_2) and pressure.

It is interesting to note that many standards also demand that measurements are corrected to a 'dry gas' basis. This simply means that the dilution due to water vapour must be removed. Gaseous concentrations are often measured 'dry' by actually removing the water vapour before analysis, but the techniques available for measuring particulates invariably measure *in situ* with water vapour included (i.e. on a 'wet gas' basis). For compliance with legislation a knowledge of the emitted water vapour is therefore often essential.

4.5 PRACTICAL LIMITATIONS

The most common source of 'errors' with continuous particulate monitors is caused by a lack of understanding of the effects of particle size variations.

Transmissometers of all types are much more sensitive to small particles than large particles. Compared with large particles, evenly spread small particles present much more surface area per unit of volume to block or scatter energy. With triboelectric probes, the situation may be completely different. Large particles will always impact upon the probe whereas smaller particles may tend to follow the gas flow.

Many have fallen into the trap of removing one or more elements from a bank of fabric filters in order to collect enough mass of particulates to establish a good calibration. As the size of particles passing through a secure filter is significantly smaller than unfiltered particles, such techniques are useless.

Similarly, if an agglomeration of smaller particles is collected during an attempt to calibrate a continuous monitor, the test could again be of little value.

Because of these frustrations and the obvious cost burden of elaborate isokinetic testing on many small plants, many regulators now only demand indicative measurement on plant with proven dust arrestment facilities.

With indicative measurement, the variations in the particulate emissions are measured, using a continuous device to indicate changes in the particulate level. No attempt is made to calibrate or quantify the emissions.

The final problems for continuous monitors are caused by condensates and by aerosols carried over from liquid scrubbers. Transmissometers have great difficulty in differentiating between liquids and solids. Triboelectric probes may be less affected, but conductive condensates can readily form a conductive bridge to the duct, and particles bound up in liquid droplets will have a different effect from dry, free particles.

Extractive systems such as those used to collect particles for beta gauges can theoretically be built to vaporize the collected liquid. Even if this can be done without blocking the extraction probes, the subsequent dilution caused by the liquid vapour can generate massive errors if it is not also measured by the analyser.

4.6 SUMMARY

By now it will be clear that no single technique for the continuous measurement of particulates is suitable for all applications. Figure 4.13 shows an attempt to summarize the limits of measurement for the four main techniques, which inevitably may lead to criticism by their manufacturers. Both the benefits and the limitations of the main techniques are summarized below.

Transmissometers

Advantages
1. Measure both smoke density and particulate density.
2. Only technique suitable for large ducts.
3. Unaffected by changes in velocity or turbulence.
4. Provides an average across the entire duct.
5. On-line automatic zero and span verification.
6. Absolute performance can be verified using standard optical filters.

Disadvantages
1. Unsuitable for very small ducts with low emissions.
2. Optical contamination can produce zero errors.

Back scatter devices

Advantages
1. Suitable for very low-level emissions.
2. Only require access from one side of the duct.
3. On-line automatic zero verification.
4. Suitable for small ducts.

Disadvantages
1. Not suitable for large ducts.
2. Off-line span verification is via a consumable aerosol.

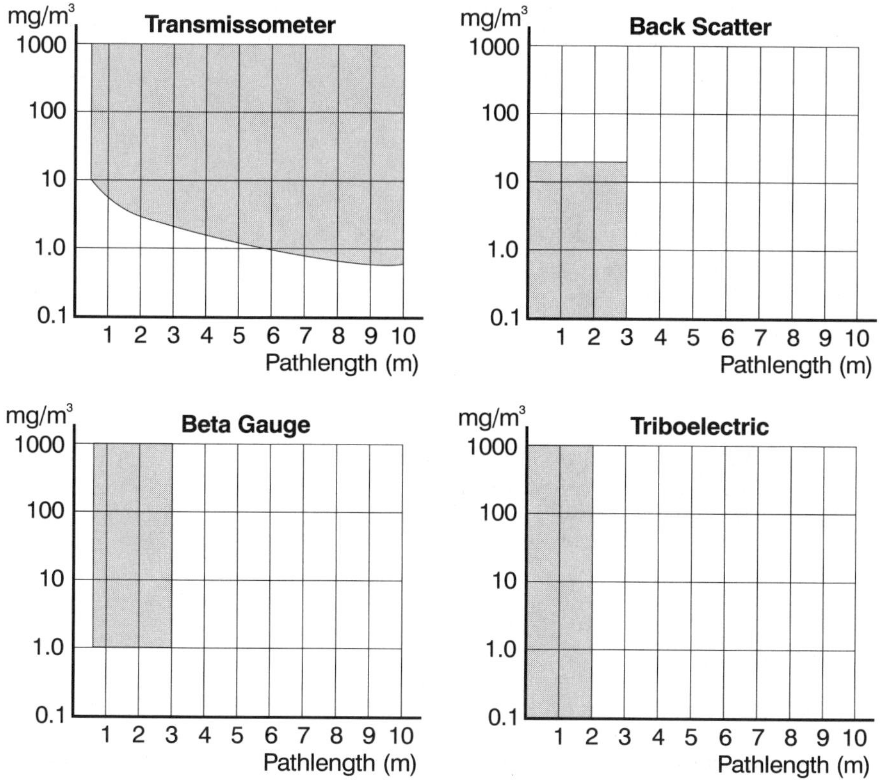

Fig. 4.13 Relative performance of continuous particulate monitors.

3. Not suitable for high-level emissions.
4. Cannot measure smoke density.

Triboelectric probes

Advantages
1. Suitable for low-level emissions.
2. Suitable for small ducts.
3. Relatively unaffected by particle build-up.
4. Relatively inexpensive.
5. Require access from only one side of the duct.

Disadvantages
1. Not suitable for large ducts.
2. Require dynamic flow measurement for mg/m^3.

3. Changes in turbulence affect output.
4. No verifiable on-line automatic zero and span check.
5. Cannot measure smoke density.

Beta gauges

Advantages
1. Independent of particle size and composition.
2. Low-level detection.
3. Suitable for small ducts.
4. On-line automatic zero verification.

Disadvantages
1. Cannot measure smoke density.
2. Sampling systems require extensive maintenance.
3. Single point analysis might not be representative.
4. Long, straight section of duct required for automatic isokinetic sampling.
5. Not suitable for large ducts.

REFERENCES

BS 2742:1969 (1991). Notes on the use of the Ringelmann and miniature smoke charts.

BS 2742, Addendum No 1:1972. The Calibration of instruments in Ringelmann Number.

Clarke, A.G. (1996) Cross duct monitoring of particulate emission by opacity fluctuations. *Environmental Technology*, 17, 101–6. [Published errata appear in *Environmental Technology*, **17**, 793 (1996).]

Lucas, D.H., Snowshill, W.L. and Cross, P.A.E. (1972) *Measurement and Control*, **15**, 9.

5

Gas sampling and conditioning

Andrew G. Clarke

5.1 GENERAL CONSIDERATIONS

Most measurements on gaseous industrial emissions are carried out by extracting a sample from the flue or chimney and transporting it either to a continuous analyser or to an absorption system for later off-line measurement. This chapter will demonstrate that gas sampling, transport and conditioning are by no means trivial operations. They can lead to significant errors in the final measurement if not undertaken correctly and the hardware represents a major part of the costs of the overall analytical system.

There are of course *in situ* analysers in which a sensor is mounted in the flue or which use cross-duct optical methods. These systems have the advantage that they avoid most of the problems associated with gas sampling. Cross-duct devices also have the advantage that they provide a signal which relates to the average concentration along the path length which may be preferable to a single point sampling system. Detailed consideration of these types of system is given in Chapters 7 and 8.

The emphasis in this chapter is on continuous emissions monitoring using sampling systems and analysers which are permanently located on site. Many of the points made will have relevance to short term campaigns using instrumental analysers (for example using mobile vans) and to batch sampling using standard reference methods involving wet chemistry. Some of the latter methods are described in Chapter 6.

The gases we wish to analyse may contain a wide variety of pollutants and other trace gas species, some of which are corrosive. They are often high in water vapour content and may have high particulate levels. They may be at elevated temperatures and pressures. In the process of withdrawing a gas sample it is necessary to ensure that:

- the sample we extract is *representative* of the gas in the flue;
- no significant chemical changes take place in the extraction and transportation process;

- the gas is brought to conditions of temperature and pressure appropriate to the analyser; and
- the analyser and sampling system are not adversely affected by unwanted particulate matter or by corrosive substances (e.g. HCl or H_2SO_4).

In the following section issues relating to representative sampling will be examined. Sections 5.3 and 5.4 cover the probes, transfer lines and sample conditioning equipment. The treatment here is very brief and for further details on individual components of sampling systems the reader should consult the suggested bibliography at the end of the chapter.

5.2 REPRESENTATIVE SAMPLING

It is essential that the sample of gas presented to an analyser is representative of the gas present in the process stream at the sampling point. This obviously has legal significance in the case of measurements undertaken to demonstrate compliance with an emissions limit. It will be argued that the gas sample on which the measurement was taken had the same concentration as the gases which were emitted and that therefore the analyser readings show beyond reasonable doubt that more than the allowable level of pollutant is being emitted from the process. There are two aspect to this issue. One is that the gas entering the end of the probe should be representative of the remainder of the gas stream in the duct or chimney. The other is that there must not be any change of composition between the probe and the analyser.

Representative sampling is best achieved by drawing the sample from a well-mixed location which might, for example, be downstream of an induced draught fan. To avoid the possibility of non-uniformity of the gas concentration it is best to sample well downstream of any point where mixing of different gas streams is occurring or where there is the possibility of air leakage into the duct. The sample will generally be taken well clear of the edge of the duct. The probe must penetrate typically, 30–50% across the duct diameter or not less than 1 m from the wall of very large ducts. For large ducts or flues, when high accuracy is required and possible non-uniformity of concentration is suspected, multipoint sampling may be necessary. The simplest approach is to use a probe which draws in gas at several different points along its length (see Chapter 9, Fig. 9.1). For highest accuracy it would entail monitoring the velocity and the gas concentration at each point and estimating a velocity-weighted average concentration, the total volume flow and hence the total emissions rate. This complex procedure, which is the norm for particulate emissions, is very rarely undertaken with gases. A compromise which may however be necessary is to carry out a preliminary survey of variations in concentration across the duct and then to locate the end of the probe at a point where typical average concentrations are to be expected.

The sample should be representative in terms of any time variations associated with the process. Obviously there are more problems characterizing the emissions from a batch chemical process than from a continuous process operating at a steady state. If one is taking grab samples then these should be sufficiently numerous to reveal the time variations. If process variations are rapid then there may be problems associated with the response time of the sampling and analysis system. This is discussed in section 5.5 below.

Sampling may be undertaken at positions where chemical reactions are continuing. This is often the case with high temperature furnace gases. Examples might include

- continuing CO and hydrocarbon oxidation in combustion systems
- NO conversion to NO_2, and SO_2 conversion to SO_3

In such situations the gases must be rapidly quenched to stop the reactions continuing in the probe and resulting in an unrepresentative gas stream reaching the analysis system. A water-cooled probe could be used or possibly an air-dilution probe. Adsorption and catalytic surface reactions in the probe must also be avoided as discussed in the next section.

5.3 SAMPLING PROBES AND TRANSFER LINES

5.3.1 Probe materials

The probe has to withstand hot, possibly wet and potentially corrosive conditions plus erosion associated with particulate impacts. Stainless steel (e.g. type 316) is the most common material used and can withstand temperatures up to 600°C. Above this temperature it may be necessary to use special alloys such as Hastelloy (up to 900°C) or ceramic materials (up to 1500°C). In some situations where cool, very wet and highly corrosive conditions exist (such as some chemical processes or downstream of a scrubber) plastic probes may be used (e.g. PVDF, up to 100°C or PTFE, up to 200°C). A stainless steel support tube may be used to protect an inner tube down which the gas passes and which is made of a material that does not react with components in the sample. H_2SO_4 vapour is best sampled using a silica-lined tube, since SO_2 to SO_3 conversion can take place on heated stainless steel. HCl is best sampled through PTFE or glass since it reacts with steel. HF is even more problematical and the options are discussed in Chapter 6.

5.3.2 Probe design

An overall view of a typical gas sampling probe is given in Fig. 5.1A. The choice of probe depends on the specific application, bearing in mind the gas temperature, gas composition, the dust loading and the particle size. The probe design depends on the precautions taken to prevent dust or droplets entering the transfer line. Various possibilities are illustrated in Fig. 5.1B, including a probe consisting of a plain tube, a probe with sintered metal pre-filter within the duct, a probe with a demister and a heated probe. External to the flue is a heated filter compartment which is illustrated in Fig. 5.1C.

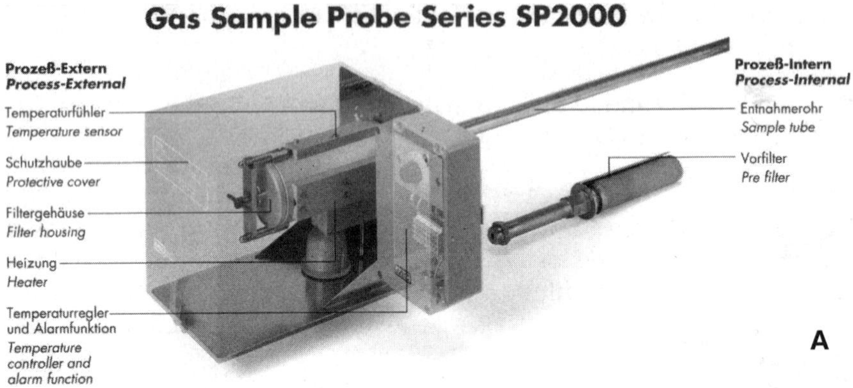

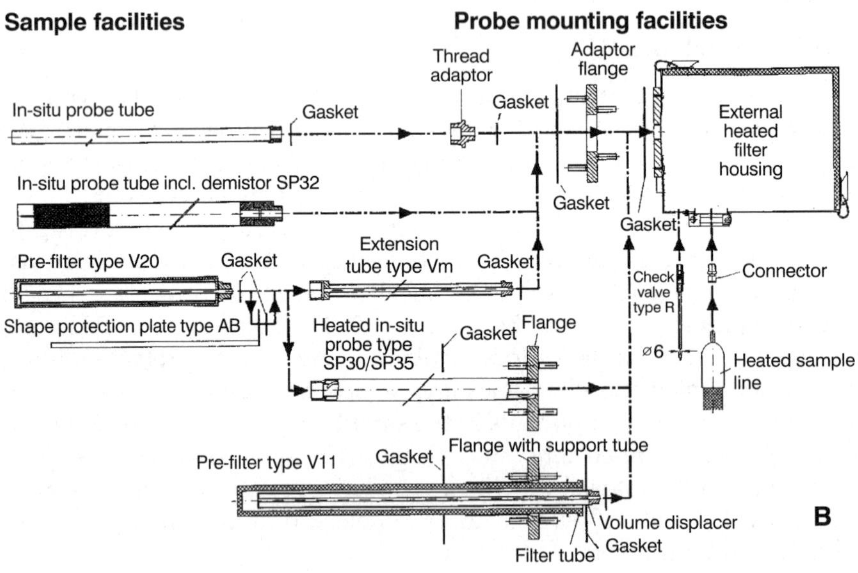

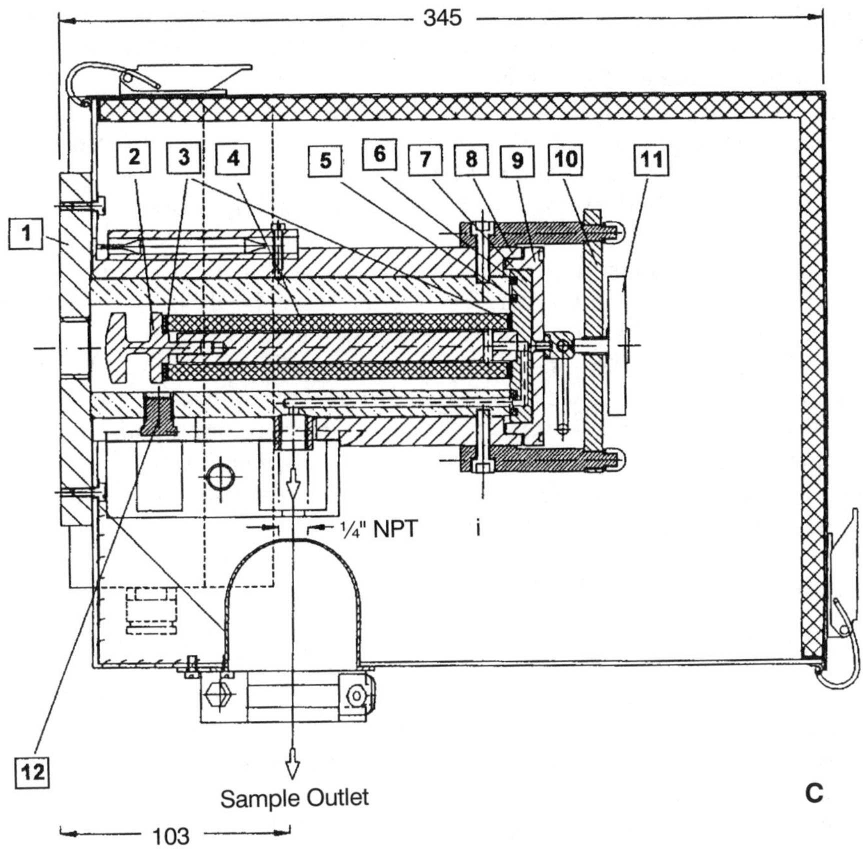

Fig. 5.1 Probe designs. (A) Overall view of a gas sampling probe. (B) Different types of sampling probes. (C) Schematic diagram of a external heated filter unit. 1, flange; 2, filter screw; 3, gasket; 4, filter element, 2 μm; 5 and 6, O-rings; 7, cover plate with filter holder; 8, aluminium heater body; 9, aluminium cover plate; 10, clamp; 11, handle screw; 12, plug. (M&C Products, courtesy of Anatrol Ltd.)

Many probes have a sintered pre-filter with fine pore size (2–5 μm) to remove particles at the gas inlet point. The upstream side of the pre-filter may need protection from erosion by a protection plate. The downstream sample conditioning system will then have a fine filter before any drying equipment. An *in situ* filter is most appropriate with high loadings of relatively coarse dust. If rapid response is important, the rather large internal volume of a pre-filter probe can be reduced with a volume displacer. The heated filter housing mounted at the end of the probe on the outside of the duct is illustrated in Fig.5.1.C. Heating is essential since condensation at the filter cannot be allowed. Filter elements may need to be able to withstand high temperatures and can be of stainless steel or sintered ceramic, down to pore size of 2 μm. A much finer pore size would introduce too high a restriction on the gas flow rate.

For continuous analysis systems both of these types of filter may become clogged and require cleaning or replacement. This may be straightforward with the external filter since it is relatively accessible. The internal pre-filter can only be reached by removing the probe from the duct. Since this is inconvenient it is possible to have a 'blow-back' or 'back-flush' system which can be triggered periodically and which clears the filter with a reverse flow of compressed air. The blow-back air should be clean, filtered and heated; otherwise, it can cause condensation in the probe.

It is highly desirable to be able to introduce calibration gas at the probe in order to be able to check for losses through the sampling system. The blow-back line can possibly be used for this purpose or an additional line to the probe is necessary. Routine calibration normally involves introduction of the gas closer to the analysers, except in the case of dilution probe systems which are discussed later.

5.3.3 Heated and cooled probes

Except when dealing with very wet gases it is not normally necessary to heat the probe itself since it rapidly adjusts to the gas temperature. In exceptional circumstances electrically heated probes may be used. Water-cooled probes are sometimes used if sampling at very high temperatures – for example, within or very close to a furnace before any heat exchange systems. The cooling serves the dual purpose of protecting the probe material (and any gas-sealing components such as gaskets) and also quenching any chemical reactions which may be taking place. Otherwise the gas composition may be evolving as the gas passes down the sampling system.

5.3.4 Transfer lines

Transfer lines are normally heated to avoid condensation. The temperature chosen depends on the gas composition:

- to avoid H_2SO_4 dew-point or hydrocarbon adsorption/condensation, 150°C is required;
- to avoid water condensation where the dew-point is unaffected by trace species the line should be at least 10° above the dew-point; 65°C would be adequate for concentrations up to 15% H_2O but if the gases are very wet arising from a scrubber or spray drier, 100°C may be required.

Transfer lines frequently involve a Teflon core surrounded by the heating coil and insulation with an outer protective sleeve. They are provided with a thermocouple adjacent to the sample line at some point and a temperature control box. A typical design is illustrated in Fig. 5.2.

There are some sampling situations where the additional expense of a heated transfer line may not be necessary. For example, if the only analyses to be undertaken are for simple combustion products CO and CO_2 plus oxygen, it does not matter if the gases are in contact with condensate. An unheated line from the probe to the sample conditioning system may be used provided that certain precautions are taken. The line must have a continuous fall in level which does not allow condensate to build-up and block the tube. A catchpot removes the liquid at the bottom of the line before the gases enter a chiller or permeation drying system. Except for short-term monitoring exercises the liquid in the catchpot will need to be continuously removed to drain or pumped away as happens in normal operation of chillers (see section 5.4.5 below). Obviously this approach cannot be used in situations where sub-zero temperatures could be encountered between the probe and the analysers since condensate might freeze and block the line.

5.4 SAMPLE CONDITIONING

5.4.1 Introduction

The sample conditioning unit draws the gas from the probe and transfer line and presents it to the analysers, cleaned, dried (not always, see

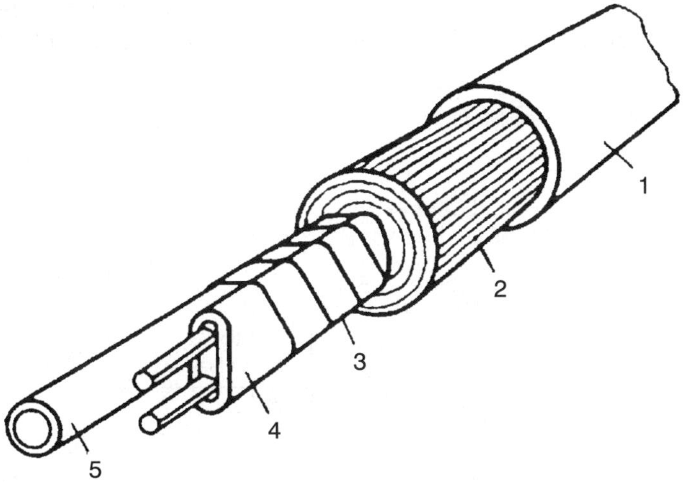

Fig. 5.2 Design of a heated transfer line. 1, PVC jacket, diameter 41–46 mm; 2, thermal insulation; 3, aluminium foil; 4, heating tape; 5, Teflon hose, 4 or 6 mm inner diameter, 6 or 8 mm outer diameter. (M&C Products, courtesy of Anatrol Ltd.)

section 5.4.2 below) and at appropriate temperature and pressure. It contains facilities to introduce zero and span calibration gases to the analysers. It must also provide for the safe venting of the gases after analysis and for safe discharge of any condensate which can be very corrosive. An example is given in Fig. 5.3.

The layout of the sampling conditioning system depends largely on the method used to handle the problem of water vapour. Typical combustion products contain water vapour at a level of about 10% by volume with a dew-point of around 46°C, although stoichiometric natural gas combustion results in very wet gas with 19% moisture and a dew-point of nearly 60°C. Other situations can produce very high moisture content with even higher dew-points, for example

- wet scrubbers
- spray driers
- steam injection to gas turbines for NO_x control
- water quench to cool gases from waste incineration

It is important that the conditioning system can accommodate the levels of water vapour which are going to be encountered. A system designed for one application may not be adequate in another. There are basically four approaches which will now be discussed.

5.4.2 Heated systems

The entire system, including analysers, is maintained at a sufficient temperature to ensure no condensation. This approach is limited by the lack

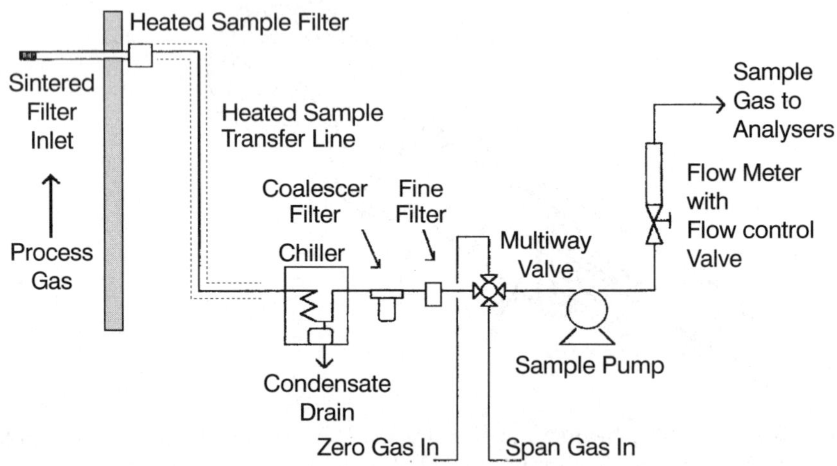

Fig. 5.3 Schematic diagram of a possible sample conditioning system.

of heated cell analysers for some components. A wet gas analysis is provided since no moisture has been removed from the gas. This means that a separate measurement of the moisture content is required in order to convert the readings to a reference dry gas basis. Ideally, a continuous moisture measurement system should be used.

Advantages

- No loss of components in condensate
- Relatively simple sample system
- No liquid to drain
- Routine maintenance limited to filter changing

Limitations

- Wet gas analysis
- Analysers and components subject to contact with corrosive gases
- Access for maintenance less straightforward since internals of analyser heated
- Limited range of high temperature analysers
- Possibility of on-going chemical reactions

Applicability

Heated flame ionization detector (FID) systems for hydrocarbons are the most commonly encountered examples. However, there are also heated UV systems for SO_2 and chemiluminescent analysers for NO_x. Heated IR systems are less common due to the problems of higher background thermal IR radiation and moisture vapour interference. Heated IR systems for HCl generally use gas filter correlation rather than traditional NDIR.

5.4.3 Dilution probe

The sample is diluted within the probe with clean, dry air by a factor of typically 100 : 1. This avoids all dew-point problems and no heated lines are required. The analysis is then carried out using low-level analysers normally used for ambient conditions. In the design shown in Fig. 5.4 the dilution air is supplied to the probe under pressure, preheated as it passes towards the front of the probe and then returns through a nozzle into an ejector. This establishes a high pressure drop in the space between the sample inlet and the ejector resulting in sample gas being drawn into the probe. The flow rate of sample is determined precisely by a critical orifice. Different orifices can be used to establish different dilution ratios between 15 : 1 to 350 : 1. The corresponding sample volumes are low compared with normal extractive systems – 500 ml/min down to 20 ml/min for the dilution ratios just quoted.

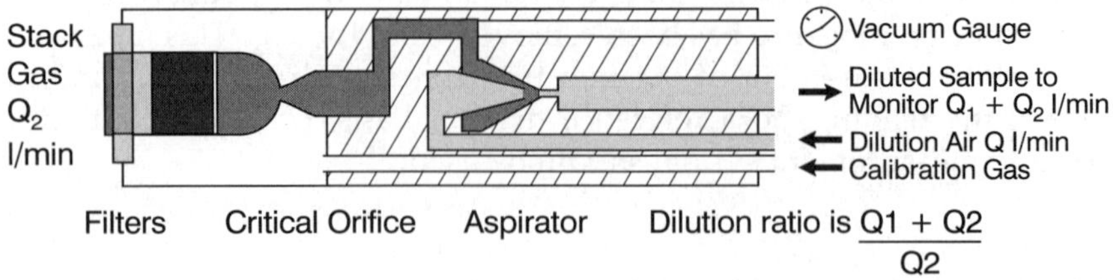

Fig. 5.4 Schematic diagram of a dilution probe. (Courtesy of ETI Group Ltd.)

Calibration is achieved by flooding the probe tip with span gas at a rate which prevents any flue gas being drawn into the system. The span gas is thus of a concentration typical of the flue gas not of the diluted sample stream. Separate calibration checks on the analysers and on the sampling system which are recommended for extractive systems (see section 5.5.2) are not required.

Filtration is achieved at the probe tip where the sample is drawn into the probe through coarse and fine filters. The low sample flow means that particle build-up on the filters is relatively slow. It is possible to use the span gas line as a means of cleaning the filters by back-flushing.

Although there is no requirement for the sample transfer line to be heated, the umbilical has to carry a minimum of three tubes for compressed air, span gas and diluted sample. A fourth tube for pressure control may also be incorporated.

The analysers, being designed for ambient levels of pollutants, may be more expensive than normal flue gas level analysers and the choice of analysis systems is more restricted. Oxygen obviously has to be measured separately and this implies a separate probe. The analysis gives the concentrations on a wet gas basis since water vapour is included in the sample. There are thus associated problems of accurate correction for water vapour which chillers and permeation drying systems avoid

Advantages

- Sampling conditioning *in situ*
- No heated lines
- Low extraction rate
- Quenches most reactions
- Problems with corrosive gases reduced

Limitations

- Wet gas analysis
- O_2 must be measured on separate system
- Requires more expensive 'ambient' analysers
- Requirement for purified, compressed dilution air

Applicability

Dilution probes can be coupled with IR analysers for gases which are strongly absorbing in the infra-red and at reasonably high concentration such as CO. However, due to the lower concentrations in the diluted sample, gas filter correlation IR may be preferred. CO_2 in air is over 300 ppm (0.03%) which gives a very high background against which to measure the diluted sample (flue gas CO_2 at 10% would give 0.2% in the diluted sample at 50 : 1 dilution ratio) although the concentration range is well within the capability of IR analysers. SO_2 and NO from combustion processes will be at too low concentration to be measured accurately by IR. SO_2 would be analysed using UV fluorescence as for the pollutant in ambient air. NO_x can be measured using chemiluminescence and total hydrocarbons by FID.

5.4.4 Permeation drier

This device relies on the unusual properties of a material called Nafion which is a copolymer of tetrafluoroethylene (Teflon) and perfluoro-3,6-dioxa-4-methyl-7-octene-sulphonic acid. In the polymer, the free sulphonic acid groups have a very high affinity for water. Each sulphonic acid group can gain up to 13 molecules of water of hydration and the polymer as a whole can gain up to 22% by weight of water. This interaction, being of a chemical nature, is rapid and reversible so water can readily permeate through a layer of the Nafion. This process is specific to water vapour and most other molecules are not absorbed. A tube with a sample gas containing high water vapour concentration on one side and a low concentration on the other side acts as a selective membrane for water vapour removal.

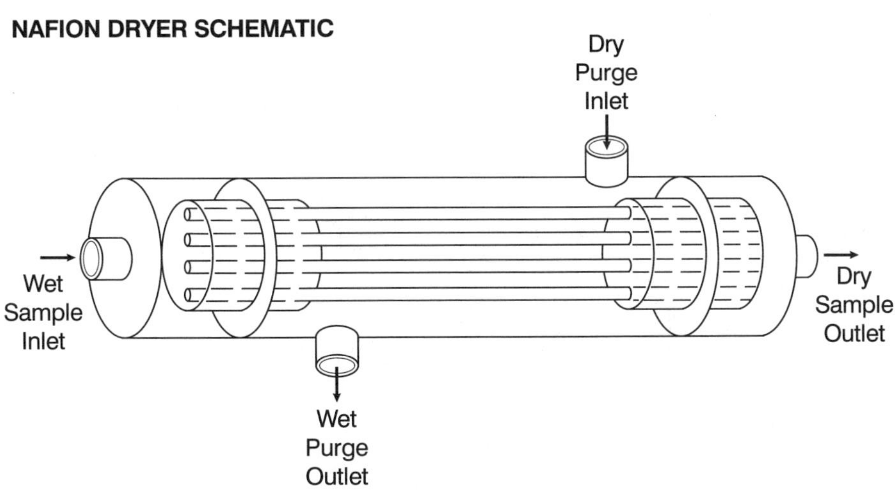

Fig. 5.5 A Nafion permeation tube drying system. (Permapure Inc., courtesy of Omnifit Ltd.)

A permeation dryer normally consists of a bundle of Nafion tubes held within a wide outer tube. Sample gas is passed through the fine tubes and a dry purge gas passed, countercurrent, through the outer tube (Fig. 5.5). Water vapour is extracted from the sample stream until the sample humidity matches that of the purge gas when equilibrium is attained. No pressure or temperature gradient is required for the process to work. A purge gas flow rate at least twice that of the sample gas is recommended. The purge gas could be filtered ambient air but this limits the extent of drying achievable to the ambient humidity level. Site compressed air may be used provided that it is filtered. Another option is to use a portion of the dried gas from the outlet side at reduced pressure as the purge gas, sometimes called the reflux approach. This arrangement is illustrated in Fig. 5.6. Provided that the reduced-pressure recycled stream has a flow rate twice that of the sample gas, the drier will operate satisfactorily. In systems involving analysers which do not change the gas composition (such as IR absorption and paramagnetic oxygen systems) all the vented gases can be returned back to the permeation tube via a pressure reducing valve – the total reflux approach.

The permeation dryer must be protected from condensates containing hydrocarbons or sulphuric acid. For example, sulphuric acid must be removed before the drier by condensation at around 80°C and elimination of the acid droplets in a coalescing filter. The permeation tube operating temperature must be above the dew-point at the sample inlet point.

Advantages

- Can be applied to most analysers
- Non-consumable

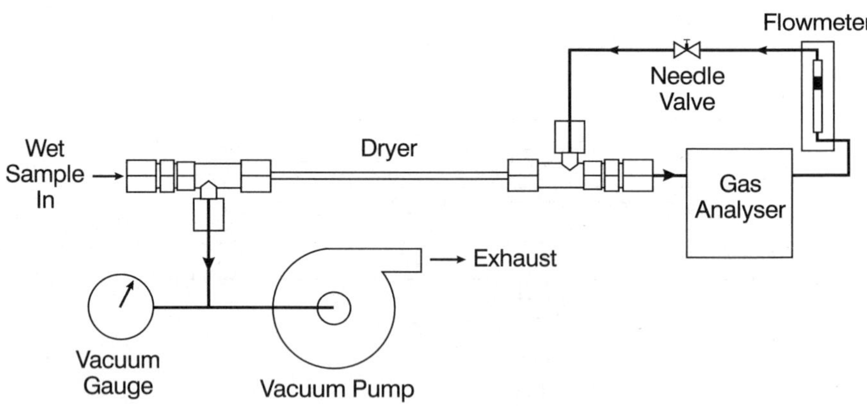

Fig. 5.6 A permeation drying system using recycled dry gas at reduced pressure as the purge gas. (Permapure Inc., courtesy of Omnifit Ltd.)

- No liquid drains
- Gives dry gas analysis
- Relatively inexpensive

Limitations
- Not suitable for some hydrocarbons
- Requires supply of dry air or added complexity of the reflux arrangement
- Trace components can condense in dryer reducing performance over time

Applicability
Permeation drying can be used for gases containing most common pollutants including H_2S, HCN and HCl. It can also be used with low-molecular weight, non-oxygenated hydrocarbons. However, it cannot be used with organic acids, alcohols and ketones, nor with ammonia and amines.

5.4.5 Chiller

The sample is rapidly chilled to around 3–5°C to remove water and acid condensate. A lower temperature could lead to potential problems of icing-up, while higher temperatures result in incomplete moisture removal. Even at 3°C the saturated vapour pressure of water corresponds to a concentration of 0.75% H_2O by volume.

Compression refrigeration or Peltier cooling systems may be used. For volumes of several litres per minute which might be required to serve a group of analysers, the sample may be split into several parallel gas streams.

Since the condensate is corrosive careful choice of materials is required. Teflon or PVC are suitable but stainless steel will be corroded. A condensate drain pump (usually a peristaltic pump) must be incorporated to ensure that the condensate is safely removed to a storage bottle or possibly reinjected into the flue. A flow of 1 litre per minute containing 10% moisture through the chiller will result in about 100 ml per day of condensate.

Chillers provide a dry gas analysis which is the reference condition normally specified in emissions limits. Some loss of water-soluble components is inevitable. This will be determined by the thermodynamic gas–liquid equilibrium which defines the solubility but also partly by the gas–liquid contact time and surface area. By good design the exposure time of the gas to the liquid surface can be minimized. This implies rapid and continuous removal of the condensate from the region where condensation is taking place.

The question arises as to whether the calibration gases should be passed *through* the chiller to the analyser (to compensate for any losses) or whether the gas *bypasses* the chiller. Different companies vary in their approaches to this problem. The calibration gas will gain a small amount of moisture vapour in the chiller coming closer to the sample gas conditions. On the other hand, the losses experienced by a dry calibration gas passing through a chiller will not be identical to those experienced by a moist sample gas suffering condensation in the chiller. Precise compensation would require the calibration gases to be prehumidified before entering the chiller, an approach which could be adopted for O_2 and CO but which would not be appropriate for more water-soluble gases such as SO_2. On balance it is therefore probably better to calibrate the analyser directly and to carry out separate checks of possible losses in the sampling system as a whole. This is discussed in Chapter 12.

Since the satisfactory operation of a chiller is critical to the sampling system and breakthrough of water vapour could lead to condensate reaching the analysers, it should be fitted with suitable alarms if either the outlet temperature rises or if the condensate drain pump fails. Some systems will automatically redirect the sample to a vent in these circumstances rather than allow contamination of analysers which could lead to an extended outage for repair.

Depending on the level of components such as sulphuric acid in the sample gas, together with the temperature/time profile of the gas during cooling a persistent acid aerosol may be formed of very fine (sub-micrometer) droplet size. This can pass right through a chiller and may not even be completely removed by a coalescing filter. A further, downstream, fine filter may then be required as shown in Fig. 5.3. In extreme cases the added expense of an electrostatic mist eliminator may be justified.

Advantages

- Dry gas analysis
- Analysers not subjected to corrosives
- Simple design
- Maintenance straightforward
- Easy to incorporate diagnostics and interlocks

Limitations

- Losses
- Not usable with very water-soluble gases (HCl, NH_3)
- Not suitable for hydrocarbons
- Liquid to drain

Applicability

Chillers are widely used for moisture removal with combustion gases containing CO, CO_2 and O_2. They obviously cannot be used for condensable hydrocarbons, nor for very water-soluble compounds such as HCl or NH_3. There may be slight losses with SO_2 and more serious losses with NO_2. There is no problem with NO, so the overall effect on the total NO_x should be small if NO is dominant in the NO/NO_2 mixture.

5.4.6 NO_2/NO converter

The analytical methods for nitrogen oxides, IR absorption and chemiluminescence, are normally used to measure NO specifically. Total NO_x (NO + NO_2) can be measured by converting the NO_2 to NO in a preliminary stage. The types of NO_x converter are described in Chapter 8. Frequently, the converter is incorporated into the analyser housing and automatic switching enables both NO, total NO_x and NO_2 (by difference) to be displayed. However, some systems will combine a simple NO analyser with a NO_x converter incorporated into the sample conditioning system. The operator can select whether to record total NO_x (with the converter in-line) or NO alone (converter by-passed).

5.4.7 Configurations involving several analysers

As seen above, different analysers have different gas sampling requirements. The configuration of the overall system must allow for this. For example consider a package involving

- heated FID for hydrocarbons
- heated IR system for HCl
- chemiluminescence for NO_x
- IR for CO and SO_2
- paramagnetic analyser for O_2

The line must be fully heated from the duct along the transfer line and past the filtration system to the take-off point for the heated FID and HCl analysers (Fig. 5.7). Thereafter the gases can be allowed to cool into a chiller or permeation system. The cool dry gas can then be distributed to the other analysers. In principle, the CO, CO_2 and O_2 analysers could be placed in series one behind the other if the flow rate requirements are compatible since these techniques are non-destructive of the pollutants and involve no addition or removal of other gases. The same is not true of chemiluminescence and nothing could be placed downstream of this analyser.

5.5 OTHER FEATURES

5.5.1 Response time

The response time of the system is the time between an instantaneous change of gas composition in the flue adjacent to the probe and the time

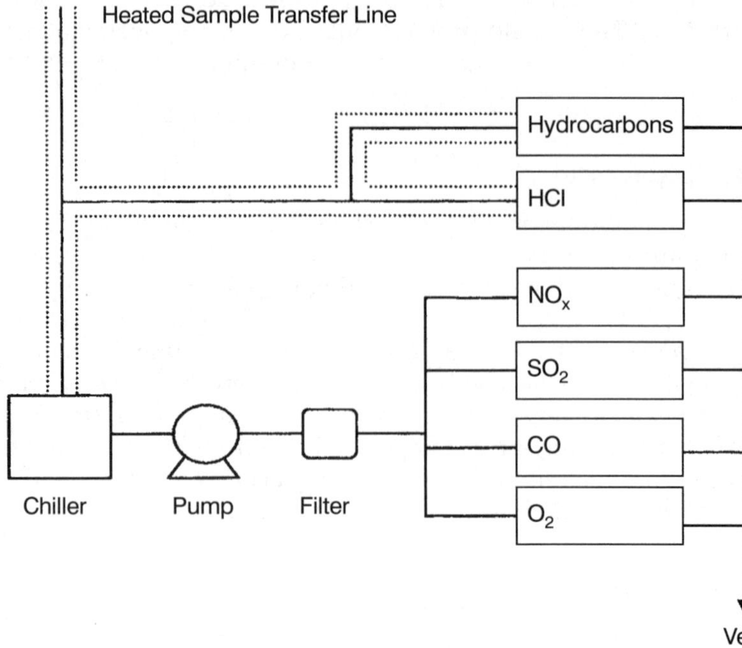

Fig. 5.7 Sample gas distribution system for multiple analysers. The heated HCl and total hydrocarbon analysers are presumed to be reliant on their own pumps and to be internally filter protected.

when that change is accurately reflected in the measured value. This time may be a few seconds or several minutes. It depends on the speed of response of the analysers themselves and on the time it takes for the whole sampling system to be flushed out and attain the new gas composition. The delay associated with the sampling system will often be longer than the delays within the analysers, so if response time is important then good design of the sampling system may be critical. If process changes which lead to gas composition changes are rapid – and if these variations need to be accurately recorded – then attention must be given to quantifying the response time. If process variations are slow or if only long-term averages (e.g. hourly or daily) are required, then this aspect of sampling can be ignored.

For sample transfer lines the average flow velocity equals the volumetric flow rate divided by the cross-sectional area. The average residence time is then the line length divided by the flow velocity. To achieve a short residence time, we want to minimize the transfer line length and also use narrow-bore tubing. In practice, 4 or 6 mm internal diameter tubing is often used. After an instantaneous change in gas composition at the inlet the outlet gas will not reach the inlet concentration after one residence time. This arises because the nature of the flow in the line is not plug flow. At typical flow rates and tube diameters the flow is laminar, which means that the velocity profile across the tube is parabolic. The peak velocity at the centre of the tube is twice the average, while the velocity at the wall is zero. To flush out the gas adjacent to the walls is therefore a slow process. It will take five residence times for the outlet concentration to reach 99% of the inlet value.

Wherever, in the sampling system, there are significant volumes of dead space there will be additional delays – examples are catchpots, chillers and coalescing filters. A poorly designed heated filter probe can also introduce a large amount of dead volume. The time taken for the sample to purge out the dead volume is called the sweep time. Sweep time is calculated by dividing the internal volume by the sample flow rate. Solution of a simple differential equation tells us that it will take 4.6 sweep times for the output to reach 99% of the inlet concentration. So, as with the flow in tubing, a factor of about 5 should be allowed in designing the system. Good design minimizes the volumes and the flow pattern so as to reduce dead volumes which are flushed out only slowly.

The overall response time will reflect the combined effect of all the delays in the system. Process changes which occur more rapidly than the time it takes for the system to reach 99% response cannot be accurately measured.

5.5.2 Sampling pump

A system involving several analysers will have a separate sampling pump drawing a greater volume than is required by the analysers. Separate pumps in the individual analysers draw the amount required from a manifold. Excess sample is vented.

The pump may be located before or after the drying system, but wherever it occurs it needs to be protected from condensate and particles by adequate filtration. Location before the chiller may require the pump to operate with gases at elevated temperature. A potential advantage is that the chiller then operates at slightly elevated pressure and it may be possible to drain condensate adequately by gravity flow and omit the peristaltic pump. With the pump downstream of the chiller the pressure is lower and a drain pump is essential.

If response time is critical a much larger pump than strictly necessary can be used to minimize residence time in the transfer line; however, the pump would then be located before the drying system and not all the gas would be dried. Increasing sample flow through the drier too much could degrade its performance.

5.5.3 Flow and pressure control

Different types of analyser have varying sensitivity to flow rate and to pressure and this will be reflected in the required control systems present in the analyser and/or sampling system. For example, the response of an IR analyser is insensitive to volume flow but directly proportional to pressure since it depends on the number of sample molecules in the IR cell. The response of a chemiluminescent analyser is quite different. This type of analyser is commonly operated at low pressure and the primary variable is the flow rate since the response depends on the number of molecules passing through the reaction chamber per second. The fine control of flow rate and pressure will normally be undertaken in the analyser however the gas conditioning system may also incorporate mass flow or pressure sensors with feedback to automatic control valves.

5.5.4 Calibration

The sample line must make provision for the introduction of calibration gases and zero gas (usually N_2) by incorporation of three-way or multi-way valves at the appropriate point (see Fig. 5.3). As discussed above, opinions differ as to whether routine day-to-day calibration should be carried out with the span and zero gases passing through a chiller if this method of sample drying is used. However, as discussed in Chapter 12, a full calibration must check the whole of the sampling system as well as the analysers. This may only be carried out infrequently but it implies injecting standard gases at the probe. One is then checking for all possible sources of error including leaks, sample loss by adsorption in sampling train, and losses by absorption in a chiller. As discussed above, with the dilution probe approach the calibration gases *have* to be injected at the probe and a separate system check is not required. Oxygen analysers generally use ambient air for calibration and this will require filtration and drying either in the sample conditioning system or at the analyser itself.

5.5.5 Effluent gas disposal

For analyser systems adjacent to the stack the effluent gases can be returned back into the flue. Otherwise they must be released from a discharge point well away from the control room and the operators. These gases are toxic and in addition may contain residual ozone from a chemiluminescent NO_x analyser (although these analysers are normally fitted with an outlet ozone scrubber).

5.5.6 Hazardous areas

In areas of a plant where flammable or explosive gases are potentially present (e.g. chemical works, refineries) all gas sampling equipment –

probes, heated filters, sample lines, chillers, etc. – must be designed to the appropriate standards of safety so as to prevent possible ignition. Hazardous areas are zoned according to the likelihood of explosive gas–air mixtures being present and specifications for the integrity of equipment set in the light of these. There are various European (CENELEC) and UK regulations (e.g. the Code of Practice in BS 5345, parts 1–9.)

FURTHER READING

Carr-Brion, K.G. and Clarke, J.R.P. (1996) *Sampling Systems for Process Analysers,* 2nd edn, Butterworth-Heinemann, Oxford, Chapter 3.

Constadine, D.M. (1985) *Process Instrumentation and Control Handbook,* 3rd edn, McGraw- Hill, Chapter 6.

Giles, J.G. (1987) Chapter 3, in *Jones' Instrument Technology, Vol. 4,* Instrumentation Systems (ed. B.E. Noltingk), Butterworths, London,

Green, D.W. (ed.) (1984) *Perry's Chemical Engineering Handbook,* 6th edn, McGraw-Hill, section 22.

6

Batch sampling and wet chemical methods of gas analysis

Andrew G. Clarke

6.1 INTRODUCTION

Much of this book deals with sophisticated optical and electronic instruments for continuous monitoring of gaseous pollutants. These have increasingly replaced traditional methods involving batch sampling and wet chemical methods of analysis. However, there are some species where continuous analysers are not available. There are also applications where results are only required periodically and therefore the additional expense of a continuous monitor is not justified. There are many industrial sites with well-equipped chemical laboratories which are quite capable of undertaking analyses of gases or liquids arising from a sampling programme relating to a flue gas or process gas stream. So there will continue to be situations where off-line analysis makes practical and economic good sense. More than this, batch sampling and subsequent analysis has been – and will continue to be – the basis of many reference test methods and for this reason, if for no other, they will continue to be used.

The scope of this chapter is very broad and it will be impossible to cover every method of analysis for every pollutant. We shall aim to cover combustion gas analysis, some aspects of gas chromatography and a range of methods involving transfer of the pollutant into aqueous solution, followed by the use of titration or instrumental techniques such as colorimetry, ion-selective electrodes, ion chromatography and spectroscopy.

The available British, German and American standard methods have been summarized in Chapter 1 and are listed in Appendices A–C. Some sampling manuals and general reference books containing detailed methodologies are listed at the end of this chapter (Harrison and Perry, 1986; Air Pollution Control Manual, 1988; HMIP, 1993; Emission Monitoring Package, 1994).

6.2 MOISTURE

The standard method of measuring moisture is to withdraw a sample through a heated line and heated filter and then to cool and collect the water with a suitable absorbent such as silica gel or with a combination of condenser plus absorption tube. The increase in weight in relation to the dry gas volume then gives the water vapour concentration. (BS 1756, Pt 4.3; US EPA, Method 4)

Given that most emission limits are given in terms of dry gas concentrations the measurement of moisture is essential in order to be able to convert any data obtained on a wet gas basis to standard conditions. Continuous emissions monitoring systems should ideally have continuous monitors, but there is no generally accepted best practice. Although there are humidity sensors on the market with a quoted temperature range of up to 150°C or even higher, most of these can only be used with clean gases and are rapidly contaminated by particles and some of the pollutant gases. A few of the approaches which can be used are described in other chapters but it may be helpful to summarize them here:

- Dual wet gas and dry gas zirconia oxygen systems.
- Heated extractive IR analysers.
- Cross duct or insertion IR analysers.
- Cross duct visible/UV analyser (e.g. DOAS)
- Impingement jet psychrometry. This German device (with TÜV approval) operates on the same principle as a wet and dry bulb thermometer, but replaces the wet bulb by a shallow pool of water onto which the gas impinges. It is unaffected by build-up of soluble pollutants.

6.3 SIMPLE COMBUSTION GASES

6.3.1 Volumetric methods

The traditional method for CO_2, CO and O_2 is the Orsat apparatus (BS 1756, Part 2). In this method, 100 ml of gas is drawn into a gas burette and the volume changes when first CO_2, then O_2 and lastly CO are removed by absorption into liquid reagents are measured. The 'CO_2' value actually includes all acid gases (e.g. SO_2 + HCl) since absorption is into KOH solution. O_2 is absorbed into alkaline pyrogallol and CO into ammoniacal cuprous chloride. A standard deviation of about 0.2% for replicate determinations is claimed. Very simple, portable, single gas analysers for CO_2 and O_2 working on the same principle of volume changes due to liquid absorption are available. It is unlikely that the accuracy of these devices is better than about ±0.5%.

A dry gas analysis is obtained from the Orsat apparatus, even though in the burette the gases are saturated with moisture at ambient temperature. All measured volumes are enhanced by the factor (1+p) where p is the saturated vapour pressure in atmospheres and this factor cancels out in calculating the percentage composition. Instead of taking a single small grab sample as just described, a better way is to collect a larger sample (e.g. 30 litres) in a flexible bag. In this way an integrated sample over a much longer sampling period can be obtained.

6.3.2 Gas chromatography methods

In this section we will cover CO_2, CO, O_2, N_2, H_2 and light hydrocarbon gases. Continuous monitoring of CO_2 and CO and CH_4 is now generally undertaken by IR absorption and either paramagnetic analysers or solid-state electrochemical probes based on zirconium oxide are used for O_2 (see Chapter 11). N_2 and H_2 are not optically active in the infra-red. Where there is a requirement for off-line analyses, gas chromatography is the most convenient technique for the species mentioned.

For these simple, permanent gases sampling is not a problem. They can be collected into Teflon sample bags or glass sample bottles. The latter are provided with a side piece covered with a rubber septum through which the sample for analysis can be withdrawn by syringe. Either direct injection onto the column or injection via a sampling loop may be employed.

For non-hydrocarbon gases a thermal conductivity detector (katharometer) is used with He or A as the carrier gas. Helium gives greater sensitivity for most gases except for H_2 in which case argon is preferable. Separation of N_2, O_2 and CO requires a molecular sieve column at about 100°C, but under these conditions CO_2 and any water vapour are retained very strongly on the column and cannot be analysed. Alternatively, one can use a porous polymeric column packing (e.g. Porapak Q) or silica gel at a similar temperature which results in appropriate retention times for CO_2 but does not resolve CO, O_2 and N_2 satisfactorily. This is a problem because either two separate GCs or single injection/dual column systems are required to analyse the full range of gases. Another approach is to use a two-in-one column (e.g. Alltech CTR) which can resolve all the components (Fig. 6.1).

The applications of thermal conductivity detectors are restricted to gases at relatively high concentrations (>0.1%, though the sensitivity depends on the combination of carrier gas and component to be analysed). Flame ionization detectors (FIDs) are far more sensitive and can go down to ppm levels, but they are restricted to organic compounds.

Light unburned hydrocarbons (methane to butane) can be sampled in a similar way to the other permanent gases and measured with a packed

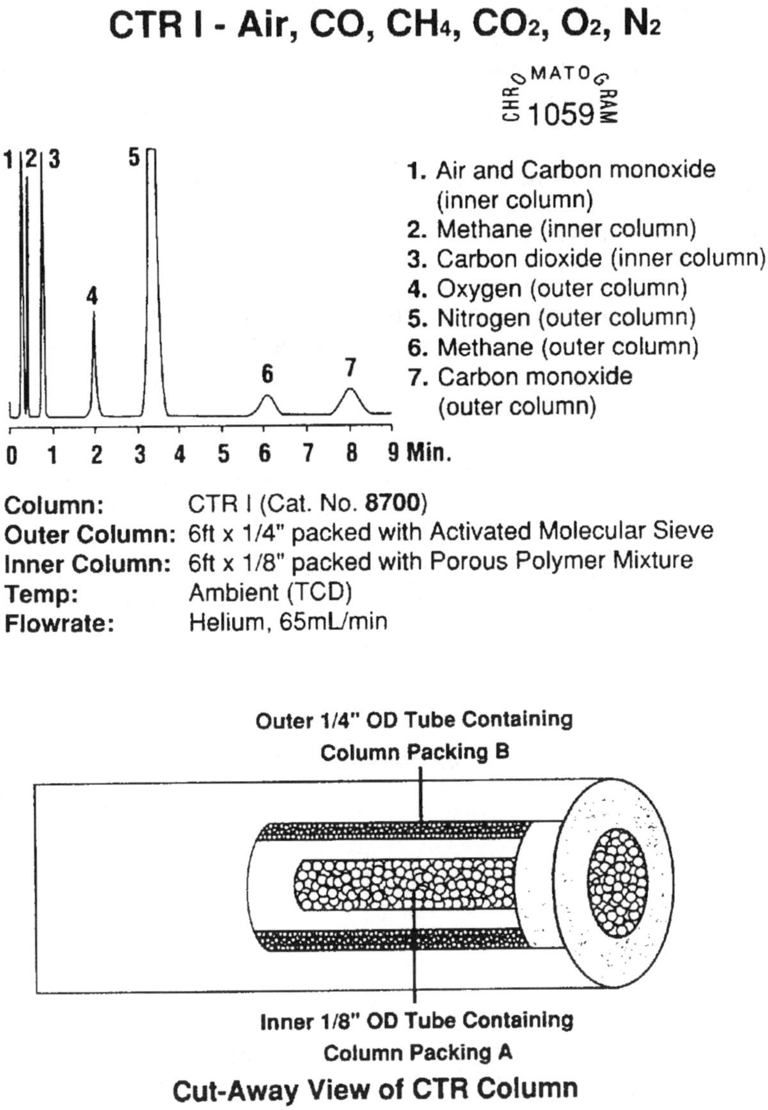

Fig. 6.1 Dual column for the analysis of simple gases by gas chromatography and corresponding chromatogram. (Courtesy of Alltech Associates.)

column and a FID detector. Isothermal conditions at temperatures just above ambient with N_2 as carrier are adequate – temperature programming is not needed. Higher hydrocarbons require greater care due to wall losses in transfer and this area is dealt with under the heading of VOCs (Chapter 9).

In this chapter we have no space to survey the applications of automated Process Gas Chromatographs which are widely used in the chemical and petroleum industries but which as yet have rarely been used for routine emissions monitoring.

6.4 OTHER GAS CHROMATOGRAPHY APPLICATIONS

6.4.1 Sulphur compounds

Sulphur, nitrogen and halogen compounds are all prescribed substances under the Environmental Protection Act. In many cases, separation and analysis by gas chromatography is quite possible. Although a FID detector can be used for all organic compounds, greater sensitivity and specificity can be achieved with flame photometric detectors FPD (for S), electron capture detectors (for halogens) and nitrogen-specific detectors. The problems associated with using gas chromatographic methods in these cases lie less with the GC technique than with the sampling. Most S, N and Cl compounds are highly polar and readily adsorb onto the walls of any transfer lines or sample containers. One may be forced to use adsorption tubes to collect the samples followed either by thermal desorption or solvent extraction to recover the compounds of interest. Such methods are covered in Chapter 9.

Kraft Pulp Mill Stack Gases

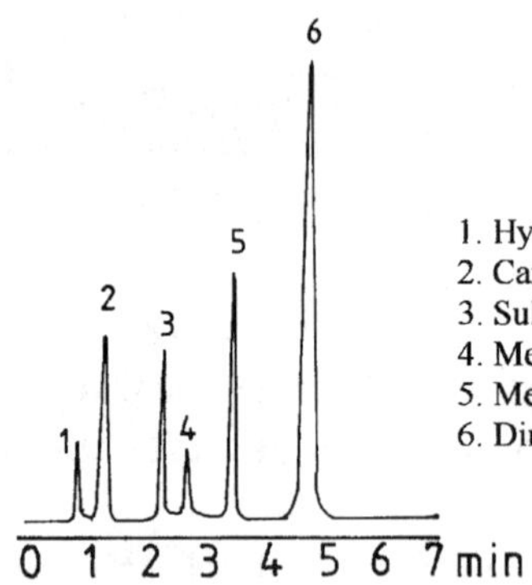

1. Hydrogen sulphide
2. Carbonyl sulphide
3. Sulphur dioxide
4. Methyl mercaptan
5. Methyl sulphide
6. Dimethyl Sulphide

Column: 30 inches x 1/8" Teflon
Packing: 18 inches Super Q, 80/100
Temp: 30°C (1 min) to 210°C at 40°C/min (FPD)
Flowrate: Helium, 40ml/min

Fig. 6.2 Sulphur-containing gases by gas chromatography. (Courtesy of Alltech Associates.)

Figure 6.2 illustrates the analysis of sulphur-containing gases from a Kraft pulp mill. Extremely inert materials are required – gas sample loops and valves should be Teflon and in the case illustrated, the column is also of Teflon. EPA Method 15 covers the analysis of hydrogen sulphide (H_2S), carbon disulphide (CS_2) and carbonyl sulphide (COS). A heated, Teflon-lined probe, Teflon filter and Teflon-lined pump are all required. After air dilution the sample is analysed by FPD/GC. A scrubber to remove SO_2 may be incorporated.

The phrase Total Reduced Sulphur (TRS) is sometimes used to describe the sum of hydrogen sulphide H_2S, methyl mercaptan CH_3SH, dimethyl sulphide $(CH_3)_2S$ and dimethyl disulphide $(CH_3)_2S_2$ normally in the context of refinery gases. The corresponding FPD/GC analysis is covered in EPA Method 16. Calibration gas mixtures cannot be used for the organic sulphides and permeation tube systems for calibration are required. An alternative approach to direct GC analysis is to oxidize the reduced sulphur compounds to SO_2 and analyse either wet chemically or by GC (EPA Methods 16A, 16B).

6.4.2 Nitrous oxide

Another application of gas chromatography is the analysis of nitrous oxide. N_2O is a greenhouse gas and there has been interest as to whether significant emissions occur as a result of combustion in addition to its predominant industrial source, the production of nylon. Initial results suggested that up to 20% of the total NO_x might be N_2O, but these high values were later shown to be due to a sampling artefact. SO_2 reacts with NO_2 in a sampling flask on the walls of which condensation has occurred and N_2O is produced. To obtain stable gas samples it is therefore necessary to remove the SO_2 and dry the gas (Muzio *et al.*, 1989). This can be done by passage through first a bubbler containing sodium hydroxide and then a drying tube (containing, for example, anhydrous calcium sulphate). The gas is analysed using an electron capture detector at 225°C with argon/5% CH_4 as carrier gas. A 2-metre Porapak Q column is used and levels of N_2O down to 1 ppm can be determined.

Continuous monitoring of N_2O can be achieved by gas filter correlation IR or conventional non-dispersive IR (NDIR) with dynamic compensation for CO and CO_2 interference.

6.5 AQUEOUS ABSORPTION AND WET CHEMICAL ANALYSIS

6.5.1 Sampling equipment

Sampling systems for continuous emissions monitoring systems are discussed in Chapter 7. The US EPA sampling train for gases involving absorption into aqueous solution is illustrated in Fig. 3.9. In this system the filter and gas absorption train are mounted just outside the duct wall on an overhead gantry. This minimizes the length of sampling line at the cost of having a greater weight of equipment to set up adjacent to the sampling point. Alternatively, it is necessary to have a heated, Teflon-lined sample line between the probe and the absorption train.

- **Probe**: the probe should be long enough to reach the centroid of smaller circular or rectangular flues or to penetrate at least 1 m into larger flues. Although stainless steel is the normal material it will be pointed out later that for certain analyses it may be necessary to use a probe lining of glass, Teflon or other material that avoids losses to the probe walls. Probe heating may be necessary to avoid condensation or adsorption problems. Water-cooled probes may be necessary at very high gas temperatures.
- **Filter**: when both gases and particles are to be measured, a heated filter box as in EPA Method 5 can be employed (see Chapter 3, Fig. 3.9). If no quantitative assessment of the particulate material is needed, then a glass-wool plug sited at the tip of the probe is adequate.
- **Collection system**: this consists of a series of bubblers or impingers to remove the pollutants into solution. Bubblers would normally have fine sintered glass frits to reduce the bubble size and ensure effective gas–liquid exchange. A typical train might consist of an initial bubbler to remove interferents, two primary bubblers/impingers and an empty bottle to act as a splash trap. If the gases entering the collection bottles are hot it is helpful to cool them in an ice box which helps to avoid evaporation problems. After the wet collection system a drying column is used to complete the removal of water vapour before metering.
- **Flow meter**: for flow rate adjustment it is convenient to have a rotameter in line, but this is not adequate for flow measurement since the flow may vary during the course of a run.
- **Total volume measurement**: normally, a dry gas meter is used downstream of the pump to ensure that the measured volume is as close to atmospheric pressure as possible. The temperature is measured at the inlet to the gas meter in order to correct the volume to standard conditions. One should also record the ambient pressure and correct the volume if it is significantly different to a standard atmosphere.
- **Flow control valve and pump**.

6.5.2 Analytical approaches

No attempt can be made in this book to go into any detail about standard analytical methods. Here, we simply list the most common approaches with minor comments as to their applicability and further information is given under the sections dealing with specific compounds.

One general point can be made and that is that *calibration* of these methods normally involves standard aqueous solutions which can be made up simply and to a high degree of accuracy. There are none of the problems associated with standard gas mixtures and the inaccuracy potentially associated with their certification.

Titration

Acid–base titrations have some application to pollutants such as SO_2 which can be converted into a strong acid in solution. They are of little value for weak acids and have the major disadvantage that they are non-specific and suffer from interference from any other acidic or basic species. Iodine chemistry offers a method for reducing agents (free I_2 reduced to I^-; example SO_2) or oxidizing agents (I^- oxidized to I_2; examples ozone, Cl_2). The amount of free iodine can be titrated after the end of the sampling period using thiosulphate. Other titration procedures make use of the colour difference between metal ion complexes with some molecules and the free species in solution (example: the thorin method for sulphate).

Colorimetry

The colour developed in solution by a suitable reaction often involving an organic dye is quantified by measuring the optical absorption in the UV or visible region of the spectrum. Colour development may be time-dependent for some applications and careful adherence to the analytical protocols is required for consistent accurate results. The colorimetric approach can be automated to produce continuous analysers (see below).

A related approach is to use filter papers or filter tape impregnated with reagents which lead to colour development. Quantification is by optical analysis of the reflected light and automated systems are commercially available for a wide variety of process applications.

Ion-selective electrodes

Ion-selective electrodes represent a relatively low-cost method of analysis for many ions in solution. Flue gas analysis for HCl, HF, HCN and NH_3 and other components can be carried out as will be described later.

The technique involves the measurement of the e.m.f. of the ion-selective electrode relative to a reference electrode. The electrode potential varies with the logarithm of the activity of the ion according to the Nernst equation and there is thus a fixed change in measured potential per decade change in activity. This is 56 mV per decade for a singly charged ion and about half this for a doubly charged ion. At very low concentrations the slope decreases and it requires considerably more care in calibration if one is operating in the non-linear region.

The activity depends on the concentration and the ionic strength of the solution. To overcome problems due to the variation of ionic strength it may be necessary to use Ionic Strength Adjustment Buffers (ISAB).

Electrodes are of various types – solid-state, plastic membrane, glass (pH and Na^+ only), gas-sensing (NH_3). As the name suggests, they are designed to be selective to the specific ion but in practice they may suffer from interferences. Table 6.1 lists electrodes which may have applications to gaseous pollution monitoring and the potential interferents. One of the most serious of these is the sulphide interference with halogen or cyanide analysis. This may require the removal of sulphide by precipitation as PbS if hydrogen sulphide is present in the original gas. There are electrodes available for various metals in addition to those listed which are less relevant to this book.

Ion chromatography

Ion chromatography has become the method of choice for most anions in solution. It is sensitive (<1 ppm in solution) and selective. There are basically two approaches. One is to use a relatively high-conductivity eluant which is suppressed after the separation column before the liquid passes into the conductivity detector (e.g. HCO_3^-/CO_3^{2-}, Dionex system). The other is to use a low-conductivity eluant which passes straight into the detector without suppression. Most sampling procedures which lead to sulphate, nitrate, chloride, etc. in aqueous solution can be determined simply and accurately by this method. Cations including ammonia and metals can also be determined if a cation exchange column instead of an anion exchange column is used.

Table 6.1 Ion-selective electrodes (Philips)

Electrode	Measurement range*		Major interferents
	ppm	mol l^{-1}	
Fluoride	0.02–1900	10^{-6}–1	OH^-
Chloride	0.35–3500	10^{-5}–0.1	CN^-, I^-, Br^-, S^{2-}
Bromide	0.08–80 000	10^{-6}–1	CN^-, I^-, S^{2-}
Iodide	0.013–127 000	10^{-7}–1	CN^-, S^{2-}
Cyanide	0.003–2600	10^{-7}–0.1	I^-, S^{2-}
Sulphide	0.32–32100	10^{-6}–0.1	Ag^+, Hg^{2+}
Nitrate	0.6–62 000	10^{-5}–1	I^-, Br, SCN^-,ClO_4^-
Ammonia (gas-sensing)	0.01–17 000	10^{-6}–1	Ionic – none

* The lower limit of the measurement range is defined in terms of the concentration at which the slope of the E versus log A curve drops below 30% of the normal value in the linear range. If it is desired to operate in the linear range to avoid calibration problems, then the lower limit should be approximately an order of magnitude higher.

Atomic absorption spectrophotometry

Atomic absorption (AA) is the most common method for the analysis of many metal ions. This will include some of those heavy metals which are required to be analysed in particulate emissions from incinerators (see Chapter 1, Table 1.1). However, not all the metals that are of concern can be analysed satisfactorily by AA. The technique also has the disadvantage that a separate lamp must be used for each element and this makes the analysis time-consuming if a large number of components are required. For these reasons emission spectroscopy is increasingly preferred.

Atomic emission spectroscopy

The limitations of AA are not shared by atomic emission spectroscopy. The particulate material is heated and vaporized by a DC or inductively coupled plasma generator, after which the optical emission spectrum as a function of wavelength is recorded. After suitable calibration all the required metallic elements can be analysed simultaneously.

6.6 SULPHUR COMPOUNDS

6.6.1 Sulphur dioxide

Sulphur dioxide is actually one of the easier compounds to measure by wet chemical techniques and there are several methods available. One is to absorb the SO_2 into hydrogen peroxide where it is oxidized to sulphuric acid. An acid–base titration can then be carried out (see BS 1756, Pt 4.5, 1977; VDI 2462/2), or the sulphate can be determined by the barium perchlorate/thorin method as described in the next section (BS 6069, Pt 4.1; EPA Method 6; VDI 2462/8). The last approach avoids interference by other acidic species (such as HCl in the case of coal combustion or incineration) which is the disadvantage of the acid–base titration. Alternatively, the sulphate can be determined by ion chromatography, although this approach is not yet reflected in standard methods. The sampling equipment envisaged in BS 6069 is very similar to that illustrated for HCl (see Fig. 6.7).

Another collection method involves iodine/potassium iodide solution which oxidizes SO_2 to sulphate while itself being reduced to colourless iodide (BS 1756, Pt 4.4; VDI 2462/1):

$$I_2 + SO_2 + H_2O \rightarrow 2I^- + 2H^+ + SO_4^{2-}$$

There are two possible sampling techniques. The easiest is to monitor the iodine coloration visually and measure the time for the colour to disappear. Because of the intensity of the light absorption by iodine (which can be enhanced by starch, to give a dark blue colour) this can be done quite simply with little inaccuracy. From the gas flow rate, the time and the total SO_2 equivalent to the iodine in the bubbler the concentration can be calculated. If it is more convenient to use a fixed sampling time one can stop the reaction before all the I_2 is used up and measure the remainder either colorimetrically or by thiosulphate titration. The difference between the iodine originally present and the measured remainder is equivalent to the SO_2 absorbed.

Interferences with the iodine method can possibly arise if other reducing agents are present. Hydrogen sulphide for example will also react with iodine.

6.6.2 Sulphuric acid vapour

A brief description of the methods for SO_2 in BS 1756, Pts 4.4 and 4.5 has been given above. The apparatus in each case allows for the separate collection of 'SO_3' – that is, sulphuric acid vapour (H_2SO_4) – and SO_2. Note that free SO_3 only occurs above about 600K and that in most flue gas situations one is actually dealing with sulphuric acid vapour.

Sampling for sulphuric acid mist or vapour from combustion processes has always been problematical. The first requirement is a heated sampling line (200°C) which ensures that the gas is maintained above the dew-point up to the point where the acid is to be condensed or absorbed. In the direct method (BS 1756, Pt 4.5) the acid is condensed and the condenser is later washed out and the washings analysed. In the other case (Pt 4.4), illustrated in Fig. 6.3, the acid is absorbed into a 4 : 1 isopropanol–water mixture. The alcohol inhibits the oxidation of SO_2 to H_2SO_4 which can occur if water alone is used. Some sulphuric acid is collected in the alcohol and some forms a mist which is removed by a sintered glass disk. The apparatus was formerly marketed in the UK but is no longer available. In the corresponding EPA Method 8, H_2SO_4 mist is collected in 80% isopropanol in an impinger and the SO_2 in two hydrogen peroxide impingers (see also VDI 2462/7).

Analysis of the sulphate in the alcohol is performed by a titration with barium perchlorate, using thorin as indicator. Barium ions associate with the sulphate. At the point when the barium is in excess there is a change in colour of the indicator. The colour change is not very distinct and care/experience is needed to obtain accurate results. It is possible to adapt the method so as to use spectrophotometry to determine the amount of barium in excess of sulphate (cf. BS 1747, part 7, 1983 where this approach is used for SO_2 absorbed from ambient air with dioxane as the solvent instead of isopropanol).

An alternative chemistry for the sulphate determination is the reaction with barium chloranilate. Absorption into isopropanol–water is again used and the barium salt of the brightly coloured chloranilate dye is insoluble in this solvent. The barium associates with the sulphate, liberating the coloured acid chloranilate which can be determined by passing the

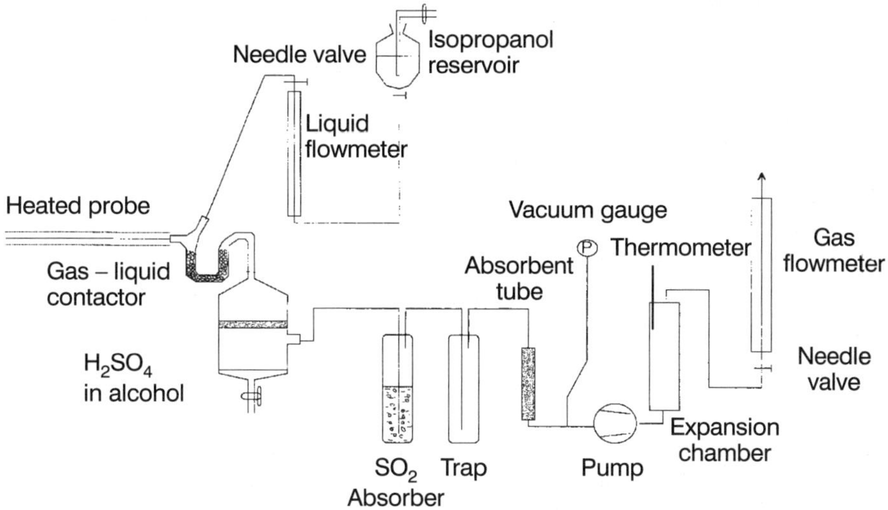

Fig. 6.3 Combined SO_2/SO_3 apparatus (BS 1756).

solution through an optical cell and measuring the increase in light absorption. A continuous analyser based on this principle was developed in the UK by the Central Electricity Generating Board and is still available commercially.

If the primary interest in sulphuric acid measurement is the potential for in-plant corrosion problems then an indirect method may be used. This is the dew-point probe described in BS 1756, part 4, section 6 and illustrated in Fig. 6.4. The principle is to measure the temperature below which a film of liquid sulphuric acid forms on a probe inserted into the duct or flue. The presence of liquid is detected by the onset of electrical conduction between two electrodes mounted on the surface of a glass thimble at the end of the probe. The surface temperature is monitored as the probe is progressively cooled by a flow of compressed air until the conductivity starts to increase. This is the dew-point temperature. In principle, it is possible to estimate the sulphuric acid vapour concentration from the dew-point temperature and the moisture vapour content of the gases from thermodynamic data. In practice, the method is best regarded as giving a semi-quantitative estimate. There are special problems if the particulate loading is high since dust also deposits on the probe.

6.6.3 Hydrogen sulphide and carbon disulphide

Gas chromatography methods for these compounds have been discussed above. Hydrogen sulphide (H_2S) can be collected by precipitation of

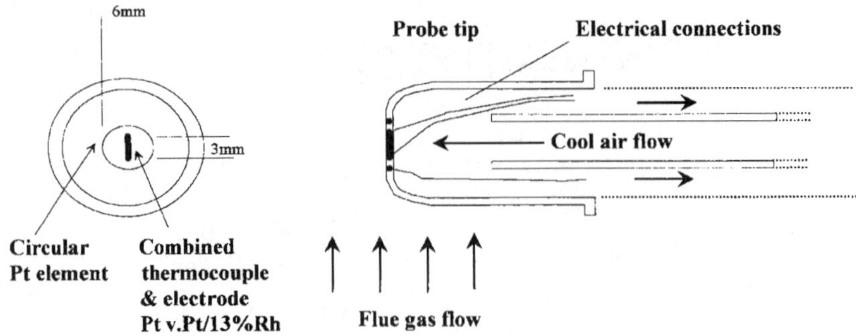

Fig. 6.4 Construction of the sulphuric acid dew-point probe (BS 1756).

cadmium sulphide from acidic solution and analysed by reaction with iodine followed by back titration of the iodine (EPA Method 11; VDI 3486/2). Prior removal of SO_2 may be necessary as this interferes. Carbon disulphide (CS_2) can also be analysed by iodimetric titration (VDI 3487/1). Alternatively, H_2S may be determined colorimetrically after collection in an alkaline suspension of cadmium hydroxide (Harrison and Perry, 1986). This precipitates CdS which is converted to the methylene blue dye for spectrophotometric determination (VDI 3486/3). Another common technique for ambient H_2S has been to measure the darkness of the stain produced on filter paper impregnated with lead acetate. The lead sulphide produced is assessed by reflectometry. Automated, continuous paper tape samplers based on this principle have been produced.

6.7 NITROGEN COMPOUNDS

6.7.1 Nitrogen oxides

The determination of NO and NO_2 by wet chemical methods is more difficult than SO_2. NO_2 is readily absorbed by aqueous solutions but gives a mixture of nitrite and nitrate. Colorimetric methods of analysis for nitrite are available (the modified Saltzman method, see BS 1756, Pt 4.9) but this leaves a fraction of the original NO_2 unaccounted for and a rather uncertain correction factor has to be applied. Oxidative absorption to nitrate followed by analysis is preferable. NO is only slightly soluble in water and must be oxidized to NO_2 before or during absorption. It is possible to promote the oxidation by passage over a solid oxidant such as chromic oxide or $MnO_2/KHSO_4$ or to pass the gas through a highly oxidizing solution such as alkaline permanganate which is the approach in EPA Method 7C. In that standard method the nitrate which is formed is reduced back to nitrite by passage over metallic cadmium before colorimetric analysis. The most commonly used method to complete the oxidation of NO to NO_2 is to allow air oxidation to occur in a collecting flask. Unfortunately, this is not a rapid process. The reaction between NO and O_2 is second-order in the NO and the reaction slows down as the concentration decreases. Chemical kinetic data suggest that 400 ppm NO + 5% O_2 should take 28.5 hours for 97.5% conversion to NO_2!

The procedure included in BS 1756, Pt 4.10 is as follows. The gas sample is admitted into an evacuated flask containing an oxidizing absorbing solution consisting of H_2O_2 in dilute sulphuric acid. The American version of this type of sampling equipment is illustrated in Fig. 6.5 (EPA Method 7). Overnight contact of the sample and solution or intermittent shaking for 2 hours is recommended followed by neutralization with NaOH and evaporation to dryness. This is to prepare a sample for nitrate analysis by the phenol disulphonic acid (PDS) method which involves nitration of the PDS in concentrated H_2SO_4 and colorimetric analysis. However, if an instrumental method for nitrate such as ion chromatography is available, the rather unpleasant chemical procedure is unnecessary. After absorption and neutralization the nitrate ion can be determined directly (EPA Method 7A) and this probably represents the best of the batch methods for NO_x. As well as the PDS method the German standards include a method based on sodium salicylate and another based on dimethyl phenol (VDI 2456 sheets 1, 8 and 10 respectively).

For nitrous oxide (N_2O), see section 6.4.2 above.

6.7.2 Ammonia and hydrogen cyanide

There is a shortage of standard methods for these compounds. The sampling approach must bear in mind that both NH_3 and HCN are polar molecules which are readily lost to the walls of the probe or transfer lines. Adsorption is not a problem, since ammonia can be captured in dilute sulphuric acid and hydrogen cyanide by dilute alkali.

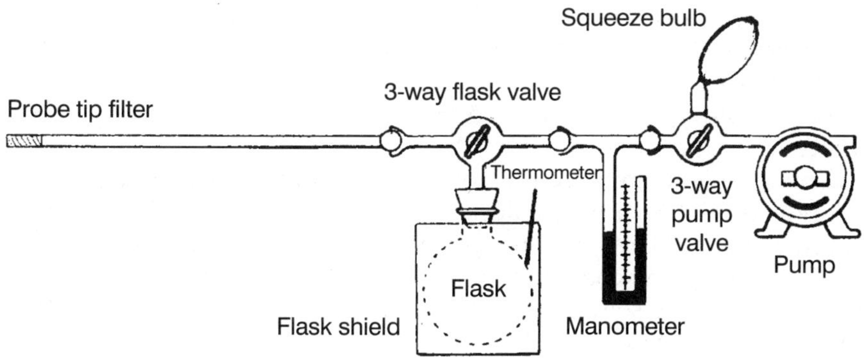

Fig. 6.5 Manual sampling of NO_x (EPA Method 7).

There are a range of colorimetric analyses for ammonium and cyanide in solution, alternatively ion chromatography or selective ion electrodes may be used.

As an example of one possible approach from work in the author's department, NH_3 and HCN were measured in the early stages of oil flames (Hampartsoumian *et al.*, 1991) where they are formed by decomposition of nitrogen-containing hydrocarbons. Their importance is as precursors to NO_x. The levels were of the order of 100 ppm or less. There were two problems associated with sampling. First, the temperature is very high and therefore a water-cooled probe was needed both to protect the probe materials and also to freeze the gas-phase reactions. Second, there was the problem of losses by surface adsorption. The gases were therefore transferred into aqueous solution *within the probe*. The sampling system is illustrated in Fig. 6.6 and is based on original work carried out at the International Flame Research Foundation at Ijmuiden (Smart and Maalman, 1987). The absorbent liquid is circulated from a reservoir into the probe by a peristaltic pump. Gas–liquid contact occurs in the probe and the liquid is returned to the reservoir where it is separated from the gas. The levels of NH_3 and HCN build up in the reservoir until a suitable level is attained for analysis. Dilute H_2SO_4 is used to collect NH_3 and dilute NaOH for HCN. The liquid is then analysed by cyanide- and gas-sensing ammonia electrodes in the laboratory. An automated continuous analyser for ammonia using a similar sampling system and ion-selective electrode analysis is commercially available. It closely resembles the HCl analyser described below.

6.8. HALOGEN COMPOUNDS

6.8.1 Hydrogen chloride

Since HCl is very water-soluble, manual sampling techniques using bubblers or impingers can be applied providing care is taken in the design of the sampling line. A heated glass probe (not stainless steel) should be used followed by a heated sample line (Teflon) and heated filter unit before the first bubbler. However, the precise method depends on the objective.

Some emission limits require 'total chloride' which includes particulate chloride as well as gaseous chloride. If this is the case, the filter unit may be omitted and the sample line kept as short as possible to avoid particulate losses. Chloride-containing particles are collected in the bubbler and are assumed to be soluble. Analysis of the solution for chloride ion then gives the sum of gaseous and particulate chloride.

Another problem is the possible presence of free chlorine (Cl_2). This will tend to be very small in coal-fired plant but can be significant in

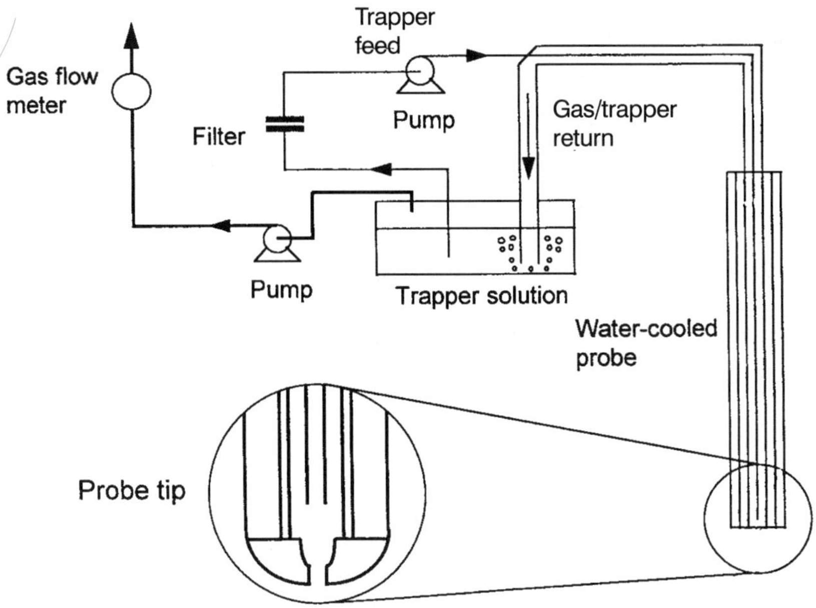

Fig. 6.6 Sampling system for NH_3 and HCN in high-temperature flames.

some incinerators. Chlorine is slightly water-soluble, producing HCl and HOCl in solution. If it is wished to measure HCl specifically then the gas should be collected in dilute sulphuric acid (EPA Method 26) in which Cl_2 is less soluble. Neutral or alkaline solutions will collect all the chlorine.

Some emissions may contain ammonium chloride which at the heated line and filter temperature will be in the vapour phase and will subsequently be collected as additional chloride in the bubbler.

The sampling equipment envisaged in the new European standard method (EN1911, 1996) for HCl is illustrated in Fig. 6.7. The probe, filter and sample line up to the first bubbler are maintained at 150°C. Absorption is into pure water and analysis in this standard is by ion chromatography or by the mercury–thiocyanate method. Ion-selective electrodes can also be used for the chloride analysis.

The instrumental analysis of HCl has proved particularly difficult due to the ease with which these gases are lost to the walls of sampling systems and the internal parts of analysers. Two approaches are to use fully heated IR systems, or to remove the gases into aqueous solution as early as possible. The latter approach is used by Severn Science Instruments coupled to a flow system with ion selective electrodes as the measurement method (Fig. 6.8)

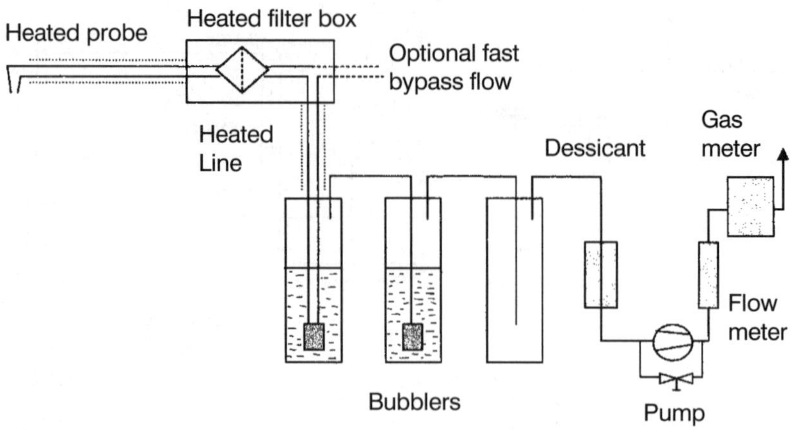

Fig. 6.7 Sampling system for HCl (European Standard EN1911). The optional fast flow bypass system is used either when combined isokinetic sampling for particles is being undertaken or if a rapid response is required necessitating a faster flow rate than can be accommodated by the bubblers.

6.8.2 Hydrogen fluoride

The problems of sampling hydrogen fluoride (HF) are even more severe than HCl. These are discussed in a recent UK ceramics industry document (Richardson and Churchill, 1992). Silica and glass will react with HF. However, this does not necessarily preclude their use since the products of the reaction include SiF_4 and H_2SiF_6 which are volatile and are collected in the bubbler where they decompose back to fluoride ions. If total fluorides are required this is satisfactory. However, if it is intended to perform an analysis to distinguish silicon-containing fluorides in the stack from HF, this approach is not satisfactory. Alumina tubes are resistant to HF attack but tend to be brittle. Of metallic materials, Monel 400 is more resistant to attack than stainless steel. The bubblers or impingers will be of inert plastic such as Teflon or polythene rather than glass. Collection is in dilute alkali (e.g. VDI 2470/1) or pure water and analysis by colorimetry (e.g. the SPADNS zirconium lake method) or ion-selective electrode (EPA methods 13A and 13B respectively). Fluoride analysis requires the used of what is called a Total Ionic Strength Adjustment Buffer (TISAB) to which the sample solution is added before measuring the electrode potential. Ion chromatography can also be used. An automated system for HF similar to that for HCl illustrated in Fig. 6.8 and using ion-selective electrode analysis is available commercially.

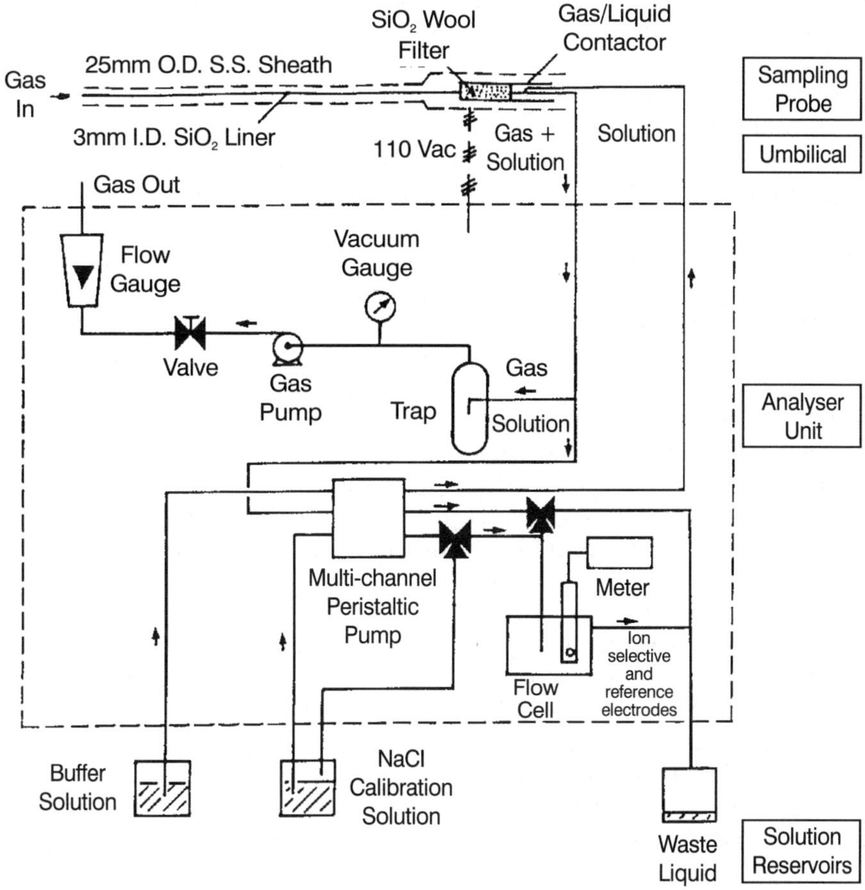

Fig. 6.8 Automated HCl analysis involving liquid scrubbing in the probe and ion-selective electrode analysis. (Courtesy Severn Science Ltd.)

6.8.3 Chlorine

See comments above on the analysis of total chloride and on the analysis of HCl/Cl_2 mixtures. Chlorine will liberate iodine from iodide solutions and this is the basis of the VDI bromide/iodide method 3488/2. Other oxidising agents will interfere with this approach. The bleaching of methyl orange is an alternative method (VDI 3488/1).

6.9 METALS

The EU Directives on incinerators require the measurement of Pb, Cr, Cu, Mn, Ni, As, Cd and Hg. German regulations also require the measurement of Thallium (Tl). The original list of US Hazardous Air Pollutants included Hg, Be and As.

The majority of the metallic compounds present in the flue will be in particulate form and will be collected on the filter. However, in some cases a small proportion may be in gaseous form – dependent on the temperature – and a filter plus back-up bubbler system may be necessary. This might apply to arsenic, antimony and selenium. Mercury compounds dissociate at combustion temperatures and the emissions are primarily in the vapour phase in the form of volatile halides and atomic mercury vapour. These can be collected in a bubbler.

The simplest method for the analysis of particulate metals is acid digestion and direct injection into a flame AA spectrometer. EPA Method 12 describes this for lead and it can also be applied to Cd, Cu and Zn. As, Sb and Se all have volatile hydrides which can be used to liberate the elements from solution and to transport them in gaseous form to the flame. (VDI 2268/2). Alternatively, the graphite furnace technique can be used (VDI 2268/4). Thallium is also measured by AA (VDI 2268/3). The gaseous mercury compounds are collected into an oxidizing solution such as 10% sulphuric acid with 1% potassium permanganate. After sampling, the permanganate is destroyed by a reducing agent. Mercuric ions are then reduced back to mercury vapour by the addition of stannous chloride and the vapour swept out of solution by a stream of air into the AA system where the absorbance at 253.7 nm is measured. A cold vapour system of this type is described in the MDHS series, no. 58 for mercury in the occupational environment.

As previously indicated, multi-element analysis is most conveniently achieved using atomic emission spectroscopy and the German standard VDI 2268/1 covers the elements Ba, Be, Cd, Co, Cr, Cu, Ni, Pb, Sr, V and Zn.

REFERENCES

Harrison, R.M. and Perry, R. (eds) (1986) *Handbook of Air Pollution Analysis*, 2nd ed, Chapman & Hall, London.

Air Pollution Control Manual of Continuous Emission Monitoring (1988) Regulations and Procedures for Emission Measurements, 2nd revised edn, German Federal Ministry for Environment, Nature Conservation and Nuclear Safety.

H.M. Inspectorate of Pollution (1993) *Monitoring Emissions of Pollutants at Source. Technical Guidance Note (Monitoring) M2*, HMSO, London.

Emission Monitoring Package (1994) Chartered Institute of Environmental Health in conjunction with National Environmental Technology Centre. Available from C.I.E.H., Chadwick House Group Ltd, Chadwick Court, 15 Hatfields, London SE1 8DJ.

Muzio, L.J., Teague, M.E., Kramlich, J.C., Cole, J.A., McCarthy, J.M. and Lyon, R.K. (1989) Errors in grab sampling measurements of N_2O from combustion sources. *J. Air Poll. Control Ass.*, **39**, 287–93.

Hampartsoumian, E., Nimmo, W., Clarke, A.G. and Williams, A. (1991) The formation of NH_3, HCN, and N_2O in an air-staged fuel oil flame. *Combustion and Flame*, **85**, 499–504.

Smart, J.P. and Maalman, T.J.F. (1987) I.F.R.F. Doc. No. F72/a/16.

European Standard EN1911 (1996) Air Quality – Stationary source emissions – manual method of determination of HCl. Part 1, sampling of gases; Part 2 Gaseous compound absorption; Part 3 Absorption solutions analysis and calculations.

Richardson, G. and Churchill, W.M. (1992) *Industry standard for the sampling of fluorides in stack gases for the ceramics industry*. Ceram Research Special Publication 135. British Ceramic Research Ltd, Stoke-on-Trent, UK.

7

Optical methods of analysis. 1 Infra-red

Colin Blackmore

7.1 INTRODUCTION

This chapter will concentrate on measuring devices utilizing the principle of the absorption of infra-red radiation by gaseous species and their subsequent use in industrial air pollution monitoring. This method – particularly the use of non-dispersive infra-red (NDIR) measurements – provides the most robust technique for the continuous monitoring of polluting gases, combining accuracy of result, together with long life of analysers and relatively low maintenance costs.

7.2 INFRA-RED SPECTRA OF TYPICAL POLLUTING GASES

For many years it has been a well-known property of heteroatomic gaseous molecules – that is those containing two or more dissimilar atoms in the molecule – to display unique absorption characteristics in the IR region of the spectrum. The intensity of the absorptions has been shown to be proportional to the quantity of the species present and independent of the intensity of the source light and a combination of these principles is given by the Beer–Lambert law (see Chapter 8).

The wavelength of light absorbed is a function of the atomic species making up the molecule and the types of vibration taking place in the molecule. It is beyond the scope of this chapter to go into detail of these vibrations but, in general, complex molecules give rise to different types of vibration, due to stretching, bending, etc. and these vibrations give characteristic absorption wavelengths, which can be quantified to give a measure of the concentration of each individual species. Analysers based on the principles of Non-Dispersive Infra-Red (NDIR), i.e. using specific wavelengths of light alone, rather than the complete spectrum, have been used for gas analysis on a commercial basis since the 1940s.

Homoatomic molecules, those containing only one type of atom within the molecule, do not give rise to characteristic vibrations which manifest themselves in the IR region of the spectrum and therefore these species cannot be measured by these techniques. In particular this applies to nitrogen and oxygen, which make up over 98% of normal air. This is particularly fortuitous since otherwise these molecules would completely mask absorptions due to industrial polluting gases, which usually exist at low levels, in the parts per million (ppm) range. The exceptions to this are normally carbon dioxide and water vapour, which may be present at quite high % levels as products of combustion.

Depending on the industrial process, from incineration, chemical works, power stations, etc., the following is a non-exhaustive list of gaseous emissions which may need to be monitored on a continuous basis:

Carbon monoxide (CO)	Nitric oxide (NO)	Ammonia (NH_3)
Carbon dioxide (CO_2)	Nitrogen dioxide (NO_2)	Sulphur dioxide (SO_2)
Oxygen (O_2)	Nitrous oxide (N_2O)	Hydrogen chloride (HCl)
Hydrocarbons (HC)	Oxides of nitrogen (NO_x)	

As stated previously, oxygen cannot be monitored using NDIR techniques and other methods must be used, typically using zirconia probes or paramagnetic analysers.

Figure 7.1 illustrates the spectrum of a pure gas, in this case carbon dioxide, subjected to radiation in the 1–15 micron (μ) wavelength region. As you will see, this gives rise to two principal absorption maxima, at 2.7 and 4.35 μ.

If the same exercise is carried out using the other gases typically present in industrial emissions, Fig. 7.2 illustrates that a characteristic 'fingerprint' is produced, by which the gas can be identified and quantified.

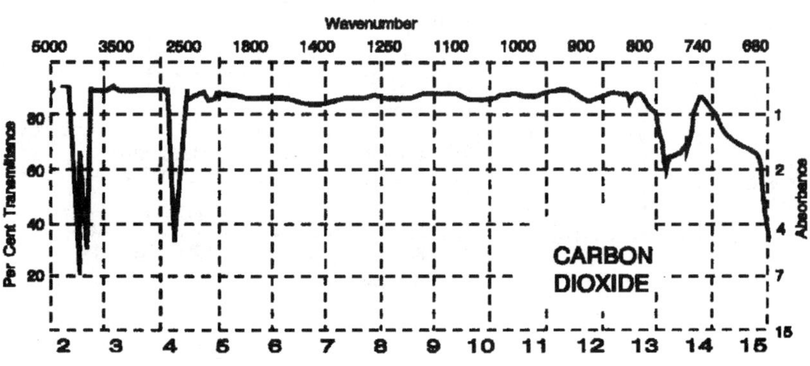

Fig. 7.1 Carbon dioxide infra-red spectrum.

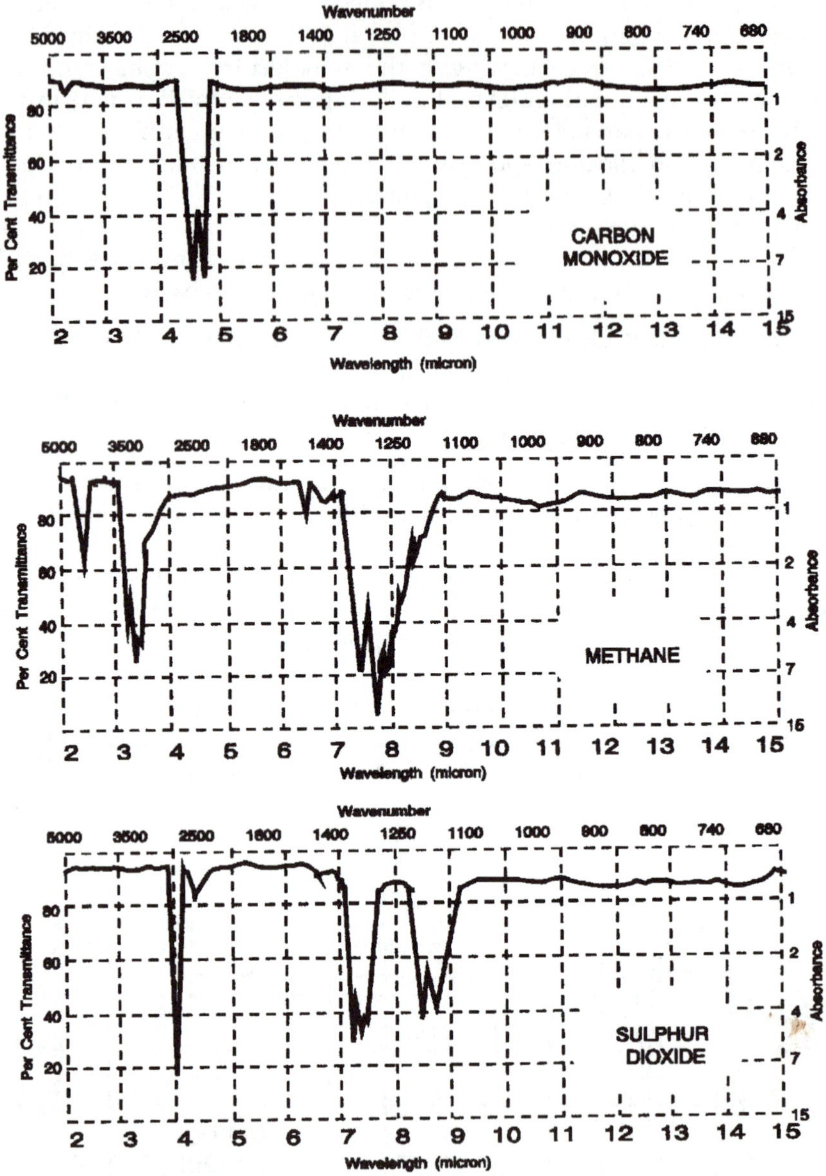

Fig. 7.2 Infra-red spectra of some common gases.

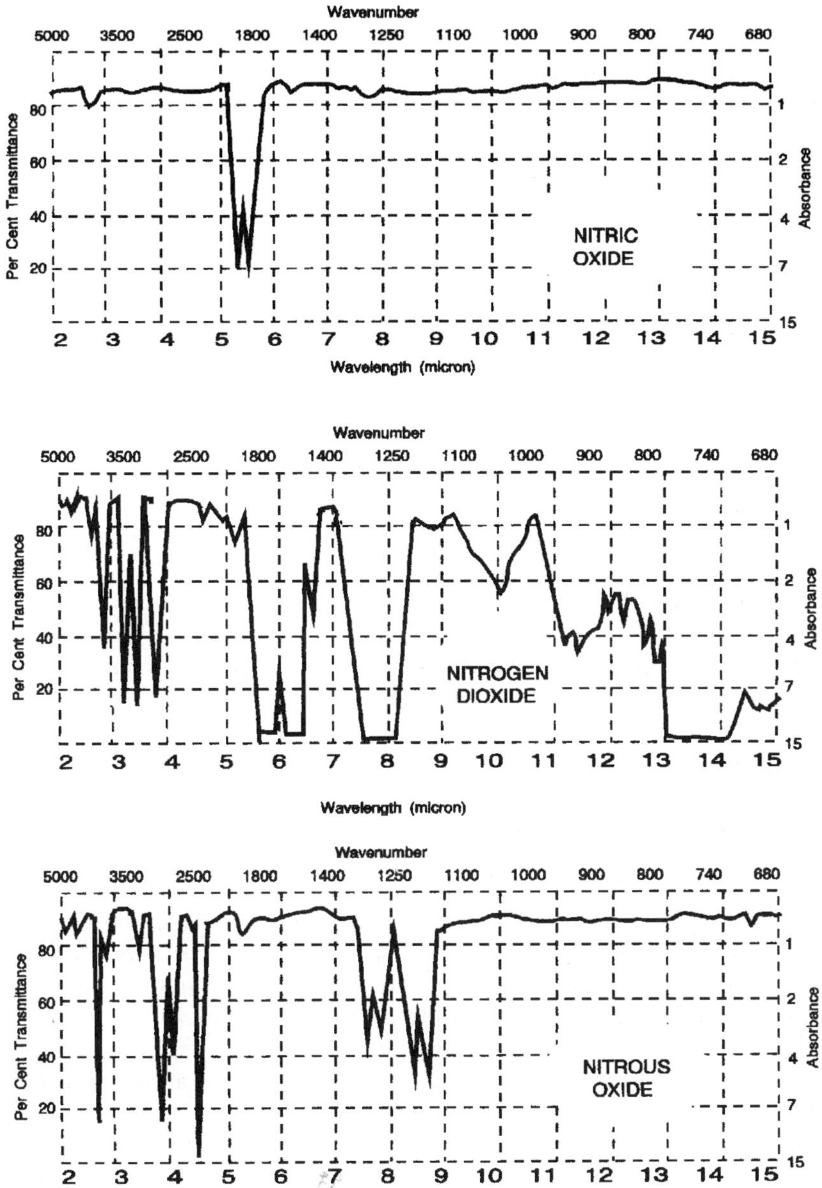

Fig. 7.2 continued

Detailed examination of the spectra for these typical products of combustion and other industrial processes shows characteristic wavelengths which may be used to allow accurate measurements to be made (Table 7.1). As mentioned before, by taking measurements of an IR-absorbing gas at a specific absorbing wavelength and comparing it with a similar measurement without the gas present, the Beer–Lambert law gives a measure of the concentration of the gas.

Table 7.1 Principal IR absorption peaks

Gas	Wavelength (micron)
CO	4.60
NO	5.30
NO_2	3.70
N_2O	4.50
CO_2	4.35
SO_2	4.00
CH_4	3.30

7.3 TYPES OF ANALYSER

7.3.1 Extractive systems

These all have the characteristic of removing the sample from the process, filtering to remove particulate matter, cooling to ambient temperature and adjusting sample pressure to atmospheric. For more details of sample preparation, see Chapter 5.

Luft analyser

This instrument is used by Analytical Development Company Limited and others. Early analysers used gas-filled detectors and one in particular, using a principle developed by Karl Luft during the 1940s, became very popular for practical measurements. This type of analyser, although still in use today, by both my company and a number of others for specific applications, was somewhat limited in use due to inherent design features, which meant that not all of the gases could be measured when present in a typical 'cocktail' of gases produced by industrial processes.

The Luft, or 'Double Beam in Space' analyser utilizes an IR light source, which is a hot wire element running at about 400°C. A chopper motor is used to modulate the source signal, together with two cells through which gas may be passed and a gas-filled detector which is basically microphonic in character and which has a diaphragm across the centre line. The latter allows any imbalance within the detector to be detected by a change in capacitance across the diaphragm. The Luft analyser is illustrated schematically in Fig. 7.3.

In a Luft-type instrument, the gas filling the detector is usually the same as the gas to be measured. This helps to fine tune the response of the detector to the detailed spectrum of the gas and helps to limit interference from other species. The reference cell normally contains nitrogen or air minus the gas to be measured and the sample gas is pumped at low pressure through the sample cell. Differences in the IR absorption in the two cells causes an imbalance in the capacitance of the electrodes on either side of the diaphragm, which is amplified to give a measure of the concentration of the gas. Further processing of the signal produces a characteristic curve, dependent on cell length and gas concentration. This curve may be linearized and resolved into Zero and Span terms, thus allowing subsequent calibration of the analyser by the introduction of Zero (not containing the species to be measured) and Span (containing the measured species at known concentration) gases and so adjusting the outputs.

Being microphonic in nature limits the suitability of the Luft-type detector for use in industrial situations. The analysers are relatively heavy to transport and are prone to vibration, so must be isolated from sources of vibration. However, these instruments offer a specification and accuracy which has resulted in their still being in daily use more than 20–25 years or more after their manufacture. Table 7.2 lists this performance and Table 7.3 shows typical minimum ranges of measurement.

In multi-gas streams there is a further limitation in that, for gases such as sulphur dioxide and nitrogen oxides, there is significant interference in

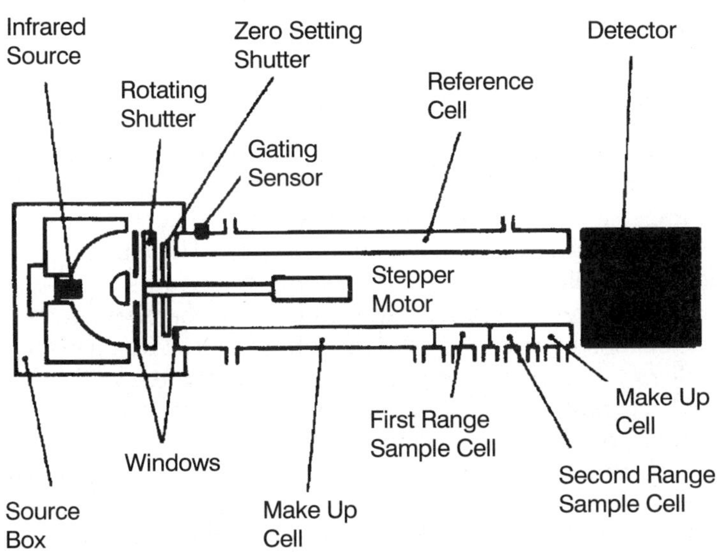

Fig. 7.3 Schematic diagram of Luft analyser.

Table 7.2 Analyser performance

Zero drift	± 0.5% FSD over 24 hours
Span stability	± 0.5% FSD over 24 hours
Linearity	± 0.5%
Detection limit	Typically <1% of FSD

Table 7.3 Lowest ranges with Luft analyser

Gas	Range (ppm)
CO	0–100
NO	0–200
NO_2	Not offered
N_2O	0–50
CO_2	0–25
SO_2	0–200
CH_4	0–200

the spectrum, particularly from water vapour. The scale of interference is such that IR analysis of these gases was found not to be acceptable by the original Luft principle in practice, in that the level of interference from water vapour (say 2% at normal temperatures and pressures) was more than the concentration of either SO_2 or NO_x present. A.D.C Limited take special precautions to limit these effects. Other manufacturers have used different methods to overcome these disadvantages, while A.D.C. Limited also developed the Rotating Filter or Gas Correlation Filter technique, described on p. 128.

Microflow sensor

A variation on the Luft principle is used by a number of manufacturers, including Rotork Analysis Limited, who utilize a microflow sensor to detect changes in the two cells or chambers. The detector consists of two chambers inter-connected by a microflow sensor. The chambers are filled with the IR-absorbing gas component of interest. During sample analysis, the pressure differential caused between the two chambers gives rise to a gas flow which is measured by the microflow sensor. The sensor consists of a couple of nickel serpentine grid resistors. When the grid resistors are heated by a current, the gas temperature distribution is symmetrical for a stationary gas, but asymmetrical for gas flowing in either direction. In the latter case, a temperature difference produced between the grid resistors induces a resistance difference which is converted into an electrical signal. Figures 7.4 and 7.5 illustrate the principle of their technique.

Another version of the microflow detector is utilized by Siemens AG. These analysers have a single cell through which the gas passes, into the

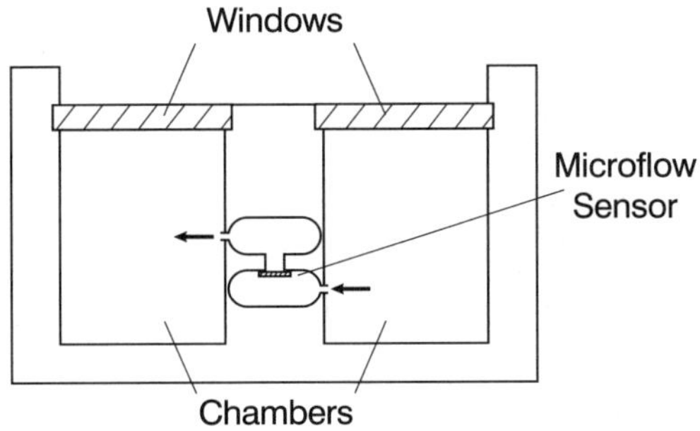

Fig. 7.4 Rotork double-beam analyser principle.

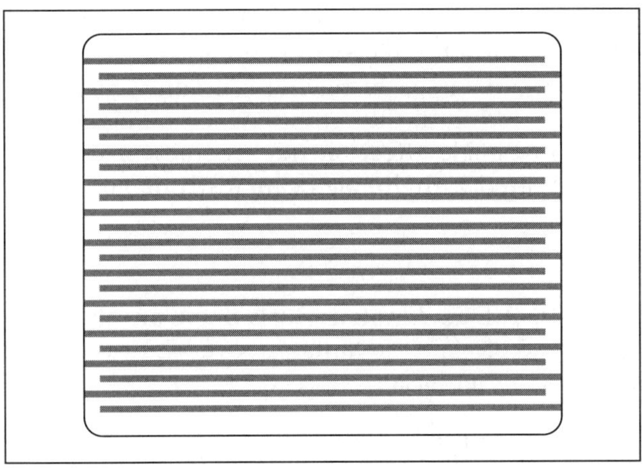

Fig. 7.5 Rotork double-beam analyser microflow sensor.

detector chamber which is divided into two gas areas, or layers (Fig. 7.6). The two areas are joined together by a capillary that contains a microflow sensor. The receiver is sensitized to the component to be measured when filled with a gas that has a spectral absorption as close as possible (or ideally identical to) the measured component.

The top detector layer absorbs the energies at the centres of the absorption bands of the sample gas; the second layer absorbs the energies of the edges. If a radiation pulse generated by the mechanical modulation reaches the detector, its absorption causes different increases in pressure in the two layers. Virtually uninfluenced by band edges or interferences, the microflow sensor detects the difference in signals. The absorption in the sample chamber increases as the concentration of the component

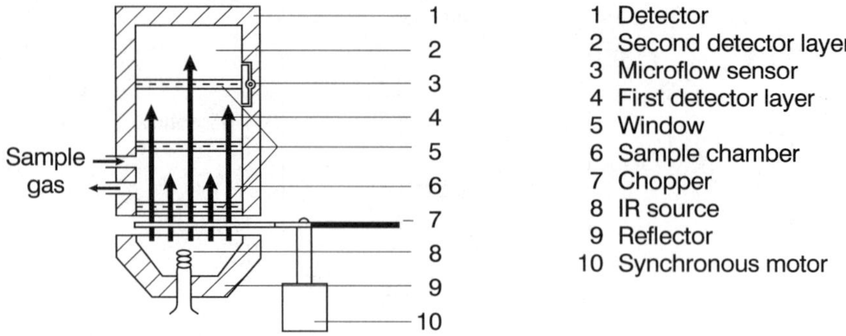

Fig. 7.6 Siemens double-beam analyser principle.

increases, and the radiation measured in the detector decreases, a measure of which is then used to characterize the detector and calibrate the analyser. The main advantage of having only one chamber is that any particulate or dust present in the sample which may adhere to the window will have a negligible effect on the accuracy of measurement and reduced drift, compared with double-beam instruments.

Gas filter correlation

In the early 1980s, A.D.C. Limited worked on an alternative type of analyser which had certain advantages over Luft-type double-beam instruments. This analyser has:

- freedom from mechanical vibration effects;
- selectivity and sensitivity at least as good as Luft instruments;
- positive discrimination against the interferences which precluded the original Luft analyser from NO/NO_x and SO_2 measurements, giving enhanced sensitivity and selectivity; and
- reduction in size and weight without loss in specification, enabling transportable systems to be constructed for continuous operation.

The analyser utilizes the gas filter correlation technique, known at A.D.C. Limited as the Rotating Filter (RF) model, which is of the 'double-beam in time' type (Figs 7.7 and 7.8).

Two optical cells are mounted in a rotating wheel to give a modulated series of signals when passed in front of a hot wire IR source. One cell is filled with the gas to be measured, while the other cell is filled with nitrogen or another gas which does not display an absorption at the same wavelength as the measurement gas. The light is collimated and passed through a thin film interference filter to limit the energy to a narrow waveband corresponding to the maximum (usually) IR absorption of the

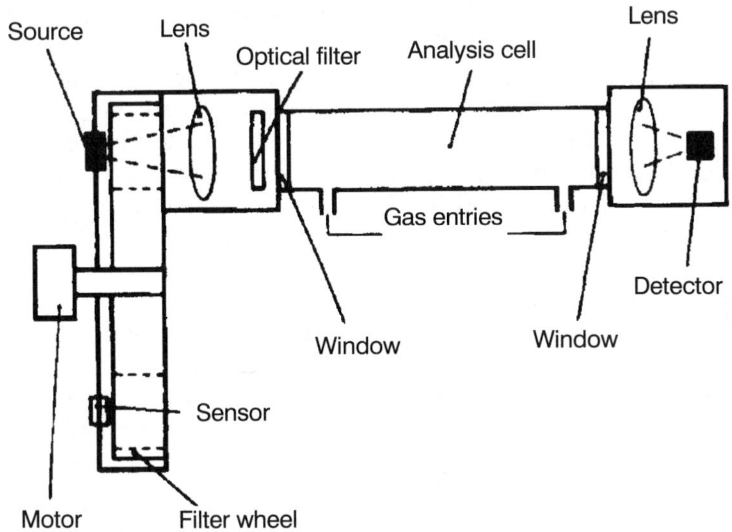

Fig. 7.7 Rotating filter analyser principle.

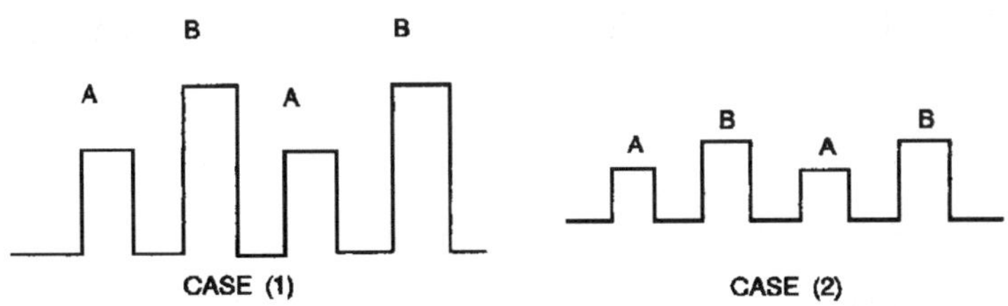

Fig. 7.8 Rotating filter analyser response.

gas. The light then passes through an analysis cell through which the sample gas is pumped and is focused onto a solid-state detector.

A practical example to illustrate the principle is a gas containing CO at the parts/million (ppm) level in a background containing CO_2 at % levels. The filter wheel cell A (shown in Fig. 7.8) will be filled with CO gas and the cell B filled with nitrogen.

When a gas passing through the analysis cell is free of CO the detector response will be as shown in CASE (1). When gas containing the absorbing gas CO enters the cell, a detector response as shown in CASE (2) is seen.

Pulse A experiences a smaller change than does pulse B, while both are reduced overall. This is because the radiation level in pulse A is already reduced by the presence of CO in the filter wheel. Amplification and linearization gives an analyser which may be calibrated over a finite range of measurements, dependent on concentration of the species present and the sample cell length.

CO_2 is also known to be present in the sample gas at % levels and has an IR absorption which would interfere with, or swamp, the CO measurement. The RF system substantially reduces the interference due to the CO_2 in the gas by tuning the detector response to the fine structure of the CO molecule, aided by the narrow waveband thin film interference filter, which thereby removes the overlap and interference due to CO_2.

Table 7.4 compares the interference effects of the original design of Luft analyser with those given by the RF series, demonstrating the greatly superior response with the RF. These principles, which are also used by a number of other manufacturers, are ideal for most combustion analyses and other industrial processes giving the following user benefits:

- freedom from noise/vibration
- single tube design overcoming any cell obscuration problems
- compact construction enabling transportable format option
- long-term stability of zero and span calibration allowing continuous operation with minimal maintenance (auto-calibration may also be fitted if required)
- high sensitivity allowing low ppm and % level measurements of most gases
- positive discrimination against interfering gases

Table 7.4 Luft versus RF analyser performance

Interfering gas		Typical Luft response	Typical RF response
Gas	Level (%)	(CO equivalent in ppm)	(CO equivalent in ppm)
CO_2	15	50–100	–1
H_2O	2	3	–0.5
CH_4	100	8–10	<1
C_2H_6	1	7	<1
NH_3	20	15	<1
NO	1	14	<1

Multi-wavelength analysers

Instead of building analysers with gas-filled cells, a number of manufacturers also use filter wheels fitted with a number of thin film interference filters. These allow the passage of light at specific wavelengths only, typically at an absorbing and a non-absorbing wavelength for the measurement of one component. The advantage of these systems is their lower cost of manufacture in most cases, but they are often limited in the lowest ranges of measurement and their specificity of gas measured, compared with the previous techniques. However, the use of more filters in the filter wheel largely overcomes these difficulties, by allowing cross-correction of the signals given at the various wavelengths (see for example, the Bruel & Kjaer Limited multi-wavelength instrument which utilizes the photo-accoustic effect). The claimed advantages of this technique over transmission spectroscopy as described previously are that:

- the signal/noise ratio is improved
- the minimum detectable limit is enhanced
- the use of a very stable transducer (microphone) reduces calibration frequency to a typical level of four times per annum
- a small volume of gas is measured (ca. 3 cm^3), enabling reduced time between samples
- the linear response enables a wide dynamic range of measurement (typically 4–5 orders of magnitude)

The inevitable consequence of these advantages however, is an increased cost, depending on the number of species to be measured, but these systems are finding increased use, particularly in very arduous conditions.

Fourier transform infra-red analysers

It is beyond the scope of this chapter to give details of Fourier transform IR (FTIR) instruments. In principle, these devices use the complete IR spectrum to compare absorbing and non-absorbing conditions, and in theory may be used to give a complete picture of the composition of a process stream. These devices are used on many occasions by the larger chemical companies in particular, but in general they are still regarded as not reliable enough techniques in most situations, requiring regular maintenance and involving a relatively high capital investment compared with other methods.

Single-beam analysers

For analyses at low ppm levels, as we have seen, the gas filter correlation or double-beam type of instruments give the best results. However, there is often a requirement for an analyser which operates at higher ranges, where the sophistication of these devices is not necessary. For example, the IR spectrum of carbon dioxide shows very strong absorptions and the levels encountered in combustion are often in the 5–15% range. For these and other high-range measurements a much simpler design of analyser can give perfectly adequate results.

These type of analysers are produced by many manufacturers, the A.D.C. Limited version being known as the single-beam (SB) analyser (see Fig. 7.9).

The limitations of this type of analyser are in the sensitivity obtainable with the instrument compared with the actual concentrations of some of the gases normally found in production situations. They also suffer from considerable cross-interference with other gases. The SB type instrument finds use in the measurement of gases in the % concentration range (Table 7.5).

7.3.2 Cross-duct systems

These all have the similarity of being mounted in a chimney stack or in ductwork near to the end of a process and are predominantly used to measure emissions to atmosphere for proving compliance with the Environmental Protection Act (1990). The principle is also applied in process control situations to an increasing extent.

Single gas measurements

As with extractive systems, cross-duct analysers (such as those supplied by Combustion Developments Company Limited, or Codel, which are described below, and Erwin Sick Optic-Electronic Ltd) utilize the absorption of IR energy in a limited band to characterize the gas concentration.

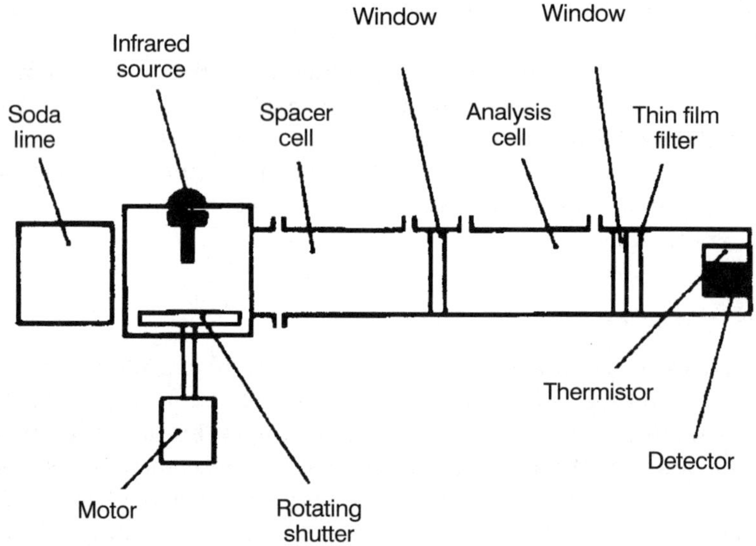

Fig. 7.9 Single beam analyser principle.

Table 7.5 Single beam analyser ranges and detection limits

Gas	Minimum Range (FSD) (%)	Detection Limit (ppm)
CO_2	0.1	10
CO	2.0	200
SO_2	1.0	100
CH_4	2.0	200
N_2O	0.1	10
NO/NO_2	Not available	Not available

So, in the case of CO a wavelength of 4.7 microns is used where absorption due to water vapour and carbon dioxide are at a minimum. By using a narrow band pass filter at wavelengths of around 4.7 micron, correctly designed CO analysers are able to ignore any possible cross-interference with water vapour and CO_2 (Fig. 7.10).

The shape of the absorption curve is fixed by the characteristics of the filter and does not change. The curve is practically flat at CO concentrations of over 10 000 ppm metres. A cross-duct monitor effectively measures CO molecules in its optical path, so the same concentration of CO will have a greater effect across a large measurement path than across a small measurement path. The term 'ppm metres' is the concentration of CO within the duct or chimney multiplied by the gas path length over which it has been measured.

Sophisticated CO analysers make two measurements of IR energy in the narrow band around 4.7 micron. Both measurements are made after

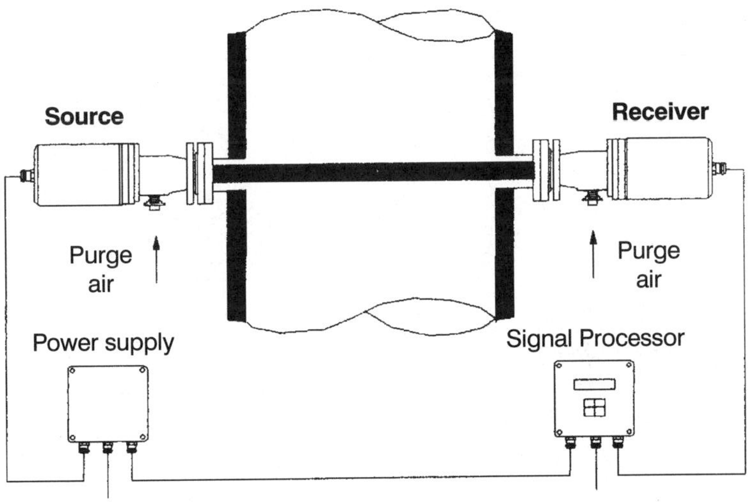

Fig. 7.10 Codel cross-duct analyser principle.

the beam has passed through the gas to be measured. Both are affected by components such as dust, which reduce the energy received from the source. One beam, however, also passes through a cell containing pure CO which absorbs all the energy capable of being absorbed by CO, providing a reference which is insensitive to any CO in the duct.

The second beam does not have such a cell in front of it and, as such, remains very sensitive to any CO present in the duct. The measurement of CO is calculated from a parameter Y, where:

$Y = 1 - K.^{D2}/_{D1}$

where D2 is the live detector output; D1 is the reference detector output; and K is a composite gain factor which takes account of all electronic and optical gains.

It is not necessary to calculate K because we know that when the constant K is correct:

Y = 0 when CO = 0.

If there is any drift in the measurement it can only be due to a change in some optical or electronic gain factor and can always be corrected by setting Y to zero when the CO level is zero. In reality, however, it is not always practical to produce a zero CO level, but if we consider the calibration curve (Fig. 7.11) we can see that:

If: Y = 0 when CO = 0
Then: Y = *a* when CO = *b*

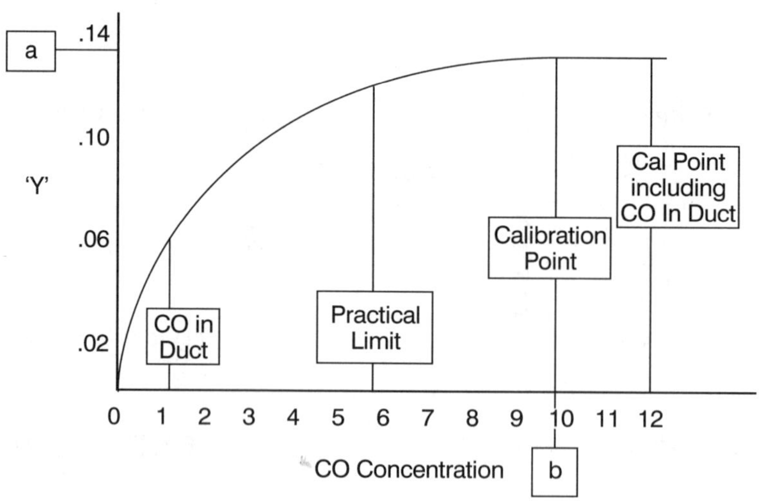

Fig. 7.11 Codel calibration curve.

We can also see from the figure that at high CO levels the parameter Y becomes completely insensitive to variable CO levels in the duct, such that Y = *a* when the CO is greater than or equal to *b*. Thus, Y = 0 when CO = 0 and all errors are eliminated.

A gas cell containing pure carbon monoxide can be introduced into the IR beam at the source. This cell represents a value of 10 000 ppm metres and provides a reference point for the calibration of the instrument. Any further CO in the duct will have a negligible effect on this reference point because the calibration curve is flat at these high concentrations of CO. Well-designed cross-duct analysers introduce this gas cell regularly (typically every few seconds) to check and if necessary modify the zero position.

This is the fundamental principle of most single gas cross-duct monitors. If they did not have this long-term stability of calibration they would be unsuitable for measurement in any real plant situation.

Multi gas measurements

A combination of the principles of the simple single gas monitor for cross-duct systems described above together with the use of gas filter correlation or multiple wavelength filter wheels described previously for extractive systems, gives the possibility of multiple gas measurement in one instrument. When cross-duct systems are mentioned, this should also be taken to include the *in situ* analysers, which contain some of the optics mounted directly in the duct or chimney stack. Manufacturers of both of these types of equipment include Erwin Sick Optic Electronic Ltd, Procal Ltd. and Codel Ltd.

7.4 COMPARISON BETWEEN EXTRACTIVE AND CROSS-DUCT SYSTEMS

Extractive systems

Advantages	**Disadvantages**
1. Cross-sensitivity very low to other species	1. Sample must be filtered
2. High accuracy achievable	2. Sample must not condense
3. Low and high ranges possible	3. Must be above dew-point
4. Low-cost hardware at stack only	4. Requires heated lines, dryers
5. Can measure NO_x, not NO only	5. Sample point selection critical
6. Easy mounting in stack	6. Relatively high maintenance
7. Maximum gas temperature 1300°C	7. Usually only one gas/analyser
8. Test gases prove entire system	8. Finite sample time delay
9. No practical duct size restriction	
10. Possibility of multiplexing	

Cross-duct systems

Advantages	**Disadvantages**
1. No sampling or extraction lines	1. Duct size, 0.8–8 m normal
2. No sample dead time	2. Max. gas temperature 400°C
3. No calibration gas required	3. Air purge to optics needed
4. Measurement over whole duct	4. Limited concentration ranges
5. Auto self-test filters	5. Lower accuracy than extractive
6. Cross-sensitivity removed by computation	6. Cross-sensitivity to some gases
7. More than one gas/analyser possible	7. Filters untraceable to standards
8. Simple and infrequent maintenance	8. Cannot measure NO_x
	9. Expensive optics on stack
	10. Alignment/vibration problem

7.5 CONCLUSIONS

This chapter is intended as a non-exhaustive description of a number of the techniques used to measure gases in practical industrial situations, using the characteristic absorption of gases in the IR region of the spectrum. All of these techniques have their place in the measurement of emissions to the atmosphere from the myriad of processes which need to be monitored now and in the future. Developments are taking place continuously to try to improve the methods, reduce the effect of conditions in which they are used (such as temperature, duct velocity, other species present) and to lower the minimum levels of measurement which are possible, with the aim of being able to meet all the future requirements of the environmental agencies.

Acknowledgements

The author would like to thank the following companies who have given permission for the reproduction of their literature in the making of this paper.

Analytical Development Company Limited, Pindar Road, Hoddesden, Herts, EN11 OAQ, UK. Tel: 01992 469638

Rotork Analysis Limited, Regal Way, Faringdon, Oxon, SN7 7BX, UK. Tel: 01367 242660

Siemens AG, Marketed by Orbital Gas Systems Ltd., Maer Lane Industrial Estate, Market Drayton, Salop, TF9 1QS, UK. Tel: 01630 658123

Bruel & Kjaer Limited, Harrow Weald Lodge, 92 Uxbridge Road, Harrow Weald, Middlesex, HA3 6BZ, UK. Tel: 0181 954 2366

Combustion Developments (CODEL) Limited, Station Building, Station Road, Bakewell, Derby, DE4 1GE, UK. Tel: 01629 814351

Erwin Sick Opto-Electronic Limited, Waldkirch House, 39 Hedley Road, St Albans, Herts, AL1 5BN, UK. Tel: 01727 831121

Procal Analytics Limited, 5 Maxwell Road, Woodston, Peterborough, Cambs., PE2 7HU, UK. Tel: 01733 232495

8

Optical methods of gas analysis. 2 Visible and ultra-violet

John Turnbull

8.1 INTRODUCTION

Gas analysis by optical methods is a wide-ranging subject with numerous techniques available for the analyst. Choosing a particular method can be a confusing and difficult choice for the inexperienced. The following rules can be applied when investigating the suitability of any given method.

1. Not all instruments using the same technique of analysis are necessarily ideally suited to that measurement.
2. The method of translating the analytical technique into an analyser has a direct effect on its performance.
3. No analyser will be suitable for measuring the full range of concentrations of any gas, i.e. from pure gas down to trace levels.
4. Instrument suitability is likely to depend on the following criteria and performance requirements:
   - The application or process in which the measurement is to be made.
   - Accuracy.
   - Repeatability.
   - Sensitivity.
   - Environment in which the measurement is to be made.

Table 8.1 lists some of the more common optical methods used in gas analysis and indicates some typical applications. The remainder of this chapter will concentrate on some of the more common optical methods found in emission monitoring today.

Table 8.1 Optical methods used in gas analysis and their application

Technique	Applications
IR spectroscopy	Emission measurements: CO, CO_2, SO_2, NO, NO_2
UV spectroscopy	Emission measurements: CO, O_2, SO_2, NO
VIS spectroscopy	Emission measurements: CO, SO_2, NO, NO_2
UV fluorescence	Ambient air HCs, SO_2, etc.
Chemiluminescence	NO_x ppb to % levels
Photoluminescence	Biological processes
Photo-ionization	CFCs, HCs etc
Colorimetry	Various liquid measurements for acid and toxic gases

8.2 ULTRA-VIOLET/VISIBLE ABSORPTION PRINCIPLES

Ultra-violet (UV)/visible absorption analysers are similar in construction and operation to infra-red (IR) analysers, but the basic principles behind their method of detection are fundamentally different:

1. Absorption of IR radiation affects the gas or sample at the molecular level, whereas UV radiation affects the orbiting electrons within the molecules of the sample.
2. Absorption of visible light may be as a result of either the infra-red process or the UV process
3. Infra-red radiation has a non-destructive effect on the sample, whereas UV radiation may break up the molecular structure of the sample.

8.2.1 IR absorption

Absorption of IR photons causes the atoms of heteroatomic molecules to vibrate. This vibration creates a changing magnetic dipole about the centre of mass which in turn causes an oscillating electric field. Where the molecule is diatomic with identical atoms there is no net change in dipole moment and therefore no energy can be absorbed. Good examples of this are oxygen, hydrogen and chlorine.

Monatomic samples such as mercury also have no IR absorption.

8.2.2 UV absorption

Absorption of UV photons excite the outermost electrons of the atoms within the molecule. The excited electrons have absorbed a UV photon in accordance with quantum theory and only specific transitions are allowed, where the electrons are pumped from a lower orbit to a higher one.

Fig. 8.1 shows a representation of the energy levels within the atom with the transition and subsequent decay to the ground state.

Only photons are absorbed where their energy level is equivalent to that in the allowable orbits for that atom/molecule. As an atom's outermost electrons are the valence electrons, the wavelengths absorbed are indicative of the molecule. The excited electrons very quickly lose the extra energy, returning to the ground state by one of the following four methods.

1. **Disassociation**. Absorption of high-energy photons can cause the electron to leave the molecule completely causing it to fragment.
2. **Re-emission**. An electron excited from the ground state by absorption of a photon can decay directly back to the ground state by the spontaneous re-emission of an identical photon. Where spontaneous re-emission occurs it is radiated isotropically and may be reabsorbed by another molecule.
3. **Fluorescence**. A molecule in a highly excited state can lose energy, by collision, to translational velocity. This may result in a transition to an intermediate level. Further spontaneous decay to the ground state will result in a photon being emitted at a lower frequency than the original absorption. A typical time for this process to occur is of the order of 1 nanosecond. UV fluorescence is used as a technique for gas analysis along with UV absorption.

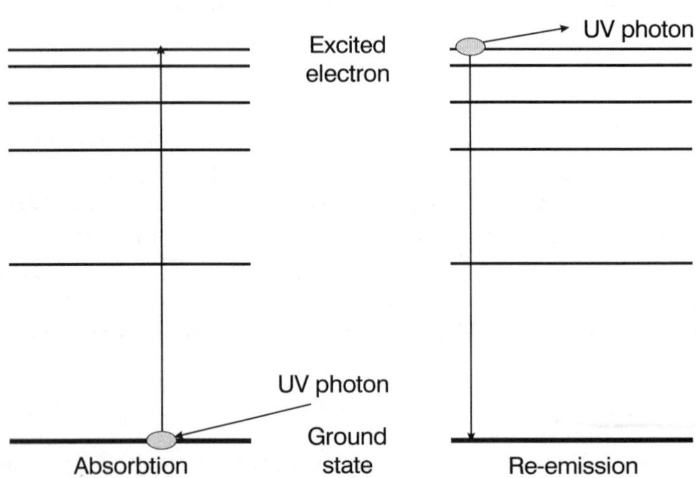

Fig. 8.1 Atomic energy levels and electronic transitions involving UV light.

4. **Phosphorescence**. The phosphorescence process is similar to fluorescence but is much delayed and emitted radiation occurs for a long time afterwards.

All of the above process occur with absorption of UV photons with electrons continually being pumped to higher energy levels and decaying by one of these processes. Figures 8.2(a–c) illustrates the actions of disassociation, re-emission and fluorescence.

8.2.3 Beer–Lambert law

Infrared, UV and visible absorption follow the Beer–Lambert Law which is an amalgamation of:

- Lambert's law, which states that the proportion of light absorbed is independent of the intensity of the source; and
- Beer's law, which states that the amount of absorption is proportional to the number of absorbing molecules.

Figure 8.3 illustrates a simple UV/visible absorption spectrometer and it can be seen that from Lambert's law, the amount of absorption or the ratio of the intensities $I_0 : I$ is independent of the absolute value of I_0.

Also, if Beer's law holds true, and the amount of absorption is proportional to the number of absorbing molecules, then the ratio $I_0 : I$ is propor-

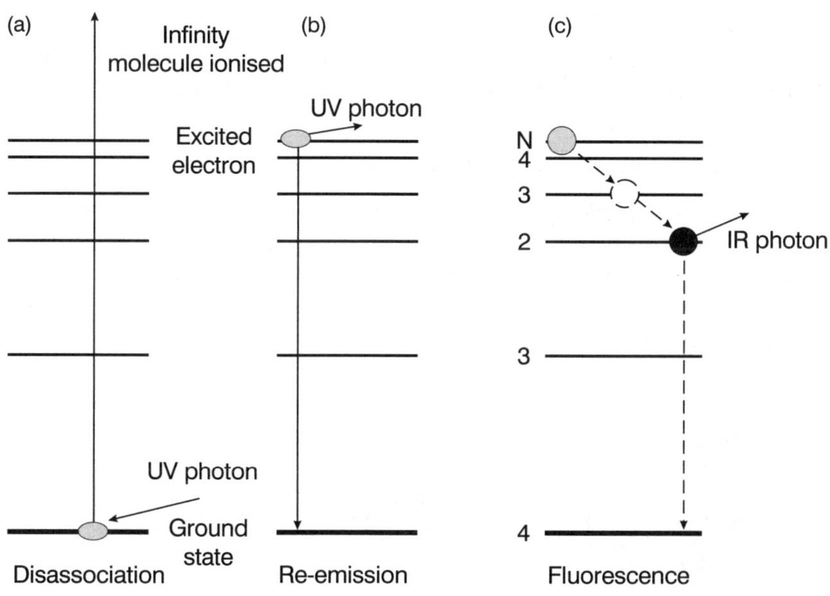

Fig. 8.2 Energy levels involved in (a) disassociation, (b) re-emission and (c) fluorescence.

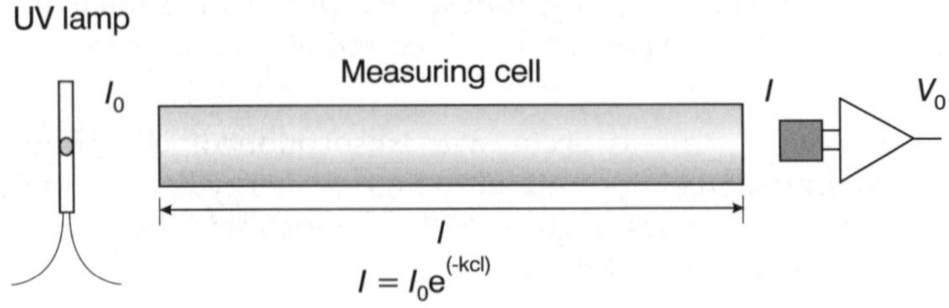

Fig. 8.3 Principle of operation of a UV/visible absorption spectrophotometer.

tional to the length of the measuring cell for a given temperature, pressure and concentration of the gas within. In practice,

$$I = I_0 \, e^{(-kcl)}$$

where

I_0 = radiation incident on cell;
I = transmitted radiation through cell;
c = concentration in sample cell;
k = conversion constant; and
l = path length through cell.

From the above,

$$\log I_0/I = Ecl$$

where $E = \mathrm{k}/\log_e 10$.

Both concentration and path length are proportional to the log of (%T) where %T is the ratio of the intensities (the transmittance).

8.3 PRACTICAL NDUV ANALYSER

In practice, the simple spectrometer illustrated in Fig. 8.3 is not practical as there is no direct measurement of the absolute intensity I_0. Consequently, there is no reference to standardized conditions. Figure 8.4 shows a double-beam instrument which resolves the need to establish an absolute value for I_0.

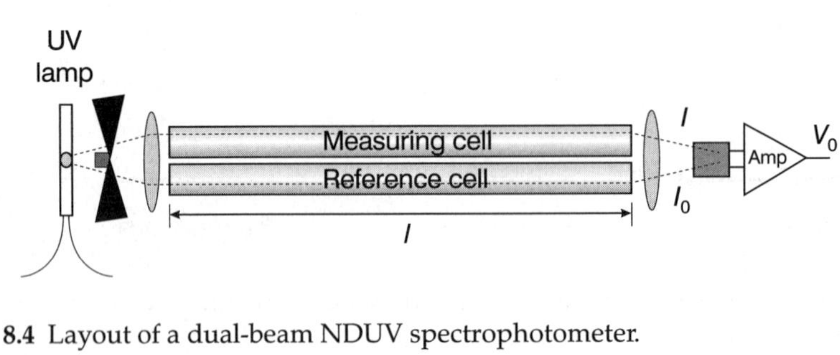

Fig. 8.4 Layout of a dual-beam NDUV spectrophotometer.

The spinning chopper disk alternately exposes the detector to the reference cell and the sample cell. The composite signal from the detector is electronically processed to produce a measurement for the values of I_0 and I.

Light from the UV lamp is collimated by the lens and passed as a parallel beam of light through both the sample and reference cells after which the objective lens re-focuses the light onto the detector. The reference cell is normally filled with a non-absorbing reference gas such as nitrogen. Filling this cell with a reference of the sample gas enables differential readings to be obtained.

Figure 8.5(a) illustrates the logarithmic output of a spectrometer for a given path length over a range of concentrations.

Figure 8.5(b) illustrates a differential output with a live zero and the signal between point A and point B is expanded electronically to produce the differential range, e.g. 98–100%.

Note: The logarithmic absorption curve is related to the path length and absolute concentration not the differential range. Expanded ranges at high concentrations can be difficult to resolve as they may be on the asymptote of the absorption curve.

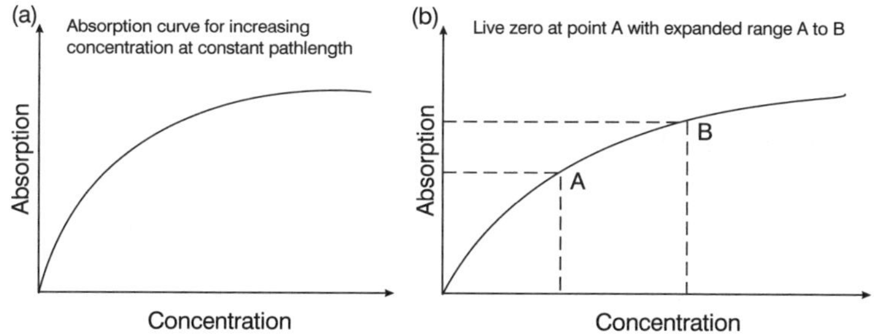

Fig. 8.5 (a) Absorption curve for increasing concentration at constant path length. (b) Use of a differential output with a live zero.

8.3.1 UV and visible lamps

As IR sources operate in the region of 800°C with relatively low energy, a Ni-chrome or tungsten wire is more than adequate. UV and visible measurement is operating at much higher photon energy levels requiring higher-powered sources. Table 8.2 lists some of the different types of UV and visible sources found in spectrophotometers. Each type has different strengths and weaknesses which affect suitability to the application. Figure 8.6 shows different types of lamp and the area of spectrum covered.

Deuterium lamp

The deuterium lamp is filled with the hydrogen isotope deuterium at low pressure. When a high voltage is applied across the anode and cathode

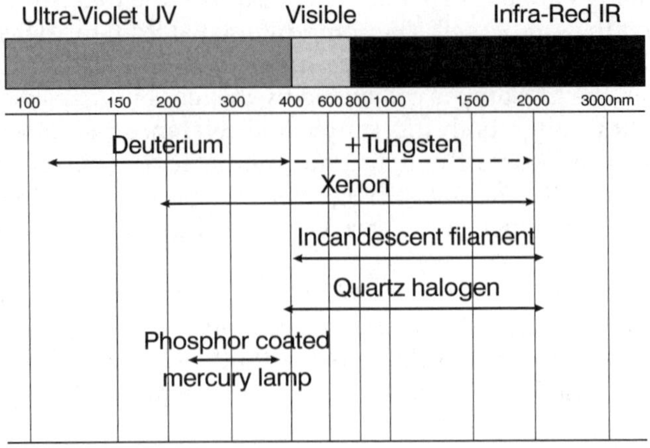

Fig. 8.6 Regions of the spectrum covered by different types of lamp.

the electrons are pumped up into a high energy gas plasma causing the lamp to 'strike', emitting broad-band UV radiation between 150 and 400 nanometres. Electrons are continually being pumped into higher orbits and transitioning down to the ground state. Differing transitions emit a different series of radiation. These are illustrated in Fig. 8.7 and show why the deuterium lamp covers a wide continuous spectrum. The high-energy deuterium lamp can have a limited lifetime as the lamp sputters the cathode onto the glass envelope reducing light emission.

Phosphor-coated mercury plasma lamp

In a simple industrial UV analyser an alternative to the deuterium lamp is the mercury plasma lamp. This lamp uses a mercury vapour and neon

Table 8.2 UV Sources

Source	Spectrum covered	Additions
Incandescent lamp	Broadband visible to near infra-red	
Quartz halogen incandescent lamp	Broad band visible to near UV	
Deuterium	Broadband UV	Addition of tungsten filament extends to IR
Xenon lamps	Broadband UV Visible and near IR	
Mercury plasma	Modified UV line spectra	Line spectra modified by phosphor to produce band emission

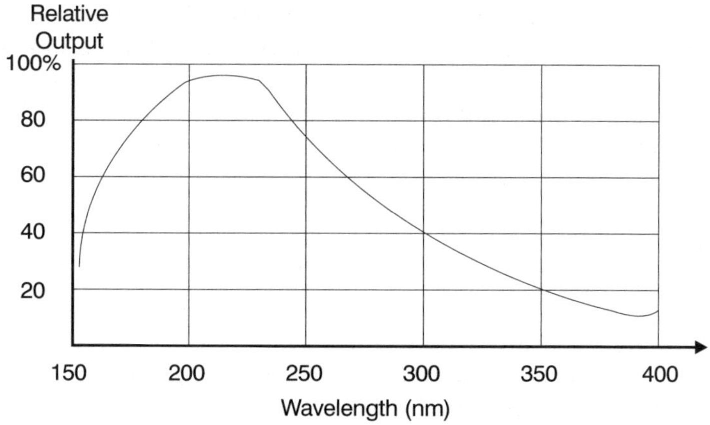

Fig. 8.7 Spectral distribution of the deuterium lamp.

plasma in the same manner as a fluorescent lamp. The mercury plasma emits a series of line spectra which in themselves would cover narrow lines in the UV spectrum. A phosphor coating absorbs the energy of the line spectra and re-emits a wider band at a different area of the UV spectrum.

Figures 8.8(a) and 8.8(b) illustrate two emitted spectra of mercury plasma lamps with different phosphors. Choosing a particular phosphor allows a dedicated analysis of specific compounds. However, note that

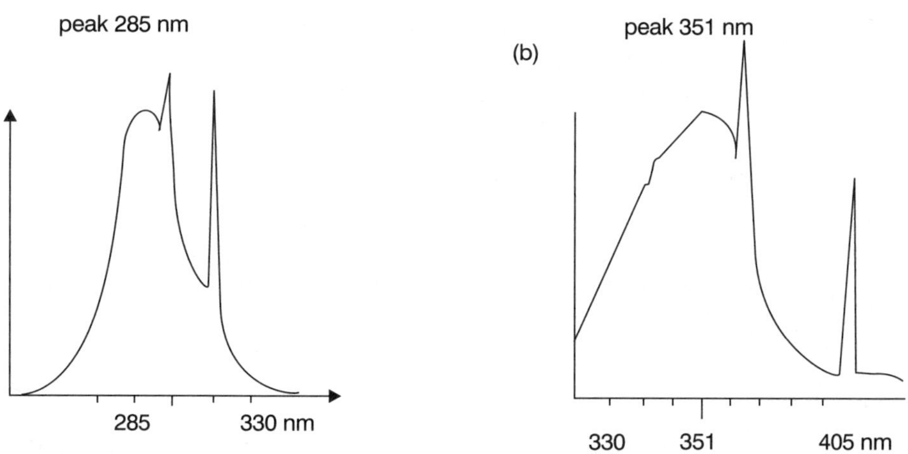

Fig. 8.8 Effect of various phosphors on the emitted spectra of mercury plasma lamps.

Fig. 8.7 shows that this type of lamp has a limited coverage of the spectrum; consequently, it is normally limited to specific compounds such as SO_2, NO_2 and Cl_2.

An advantage to this type of lamp is the long lifetime, similar to that of the fluorescent lamp.

Incandescent lamp

There are two types of incandescent lamp in common use – the filament lamp and the quartz halogen lamp.

- Filament lamps mainly emit in the visible region of the spectrum and are of little use to UV analysers. However, they do extend through the visible and into the infra-red.
- Quartz halogen lamps are a variation on the filament lamp which run at a higher temperature, enabling them to emit a small amount in the near UV. This is of limited use in the UV, but can be used for the measurement of NO_2.

Xenon lamps

Xenon lamps are used in open path spectrometers such as the Opsis system. These lamps provide wide coverage of the spectrum from approximately 180 to 2000 nanometres. The wide spectral emission takes in the far-UV, near-UV, visible, near-IR, and part of the mid-IR.

8.3.2 Light chopper

The light chopper interrupts the radiation from the source, enabling a phase-sensitive detector and discrimination circuits to compensate for random noise, temperature drift in detector, etc.

8.3.3 Detectors

There are many detectors available for UV and visible spectrometers. The higher-energy photons are relatively easy to detect compared with infrared. Typical detectors are photo-voltaic diodes, photo-resistive diodes, photomultiplier tubes, etc.

8.3.4 Filters

NDUV is an acronym for non-dispersive ultra-violet, meaning that radiation is not separated into wavelengths by dispersion prisms, diffraction gratings, etc. The whole spectrum emitted by the lamp would therefore reach the detector, swamping it compared with the small amount absorbed by gas in the sample cell. This may lead to poor sensitivity and high cross-sensitivity in a broad-band detector. Filters sensitizing the detector to a narrow band of interest reduce this problem.

8.3.5 UV spectra and typical measurements by UV

Table 8.3 and Fig. 8.9 identify some of the common gas measurements made by UV absorption. Some of these gases are highly reactive and difficult to calibrate and many of them are not positioned favourably in the spectrum to take advantage of long-life sources such as the mercury plasma lamp. However, one of the main strengths of UV absorption is the absence of water vapour effects. Consequently, a great advantage is gained when making sensitive measurements of gases containing water vapour. Also indicated are the dissociation points of the SO_2, NO and NO_2 spectra.

Table 8.3

Gases measured at UV wavelengths	Wavelength (nm)
Sulphur dioxide	285
Ammonia	200
Hydrogen sulphide	200–230
Oxygen/ozone	170
Chlorine	280–380
Nitrogen dioxide	380–420

8.4 DISPERSIVE UV

NDUV analysers are normally dedicated to one specific measurement by the selection of the optical interference filter. Using a broad-band source and several optical interference filters it is possible to produce a multi-component analyser where the filters are mounted on a rotating wheel, sequentially sensitizing the analyser to different wavelengths.

Dispersive analysers achieve the same effect by using a monochromator to select the desired wavelengths in the spectrum. Separation is achieved by rotation of a dispersion prism, diffraction grating or both in the light beam. Figure 8.10 illustrates a typical dispersion spectrometer which is commonly found in both laboratories and industries.

8.4.1 Diode array spectrophotometer

The diode array spectrophotometer is a variation on that in Figure 8.10. However, instead of rotating the prism or diffraction to present a narrow band of the spectrum to a single detector, an array of photodiodes monitor

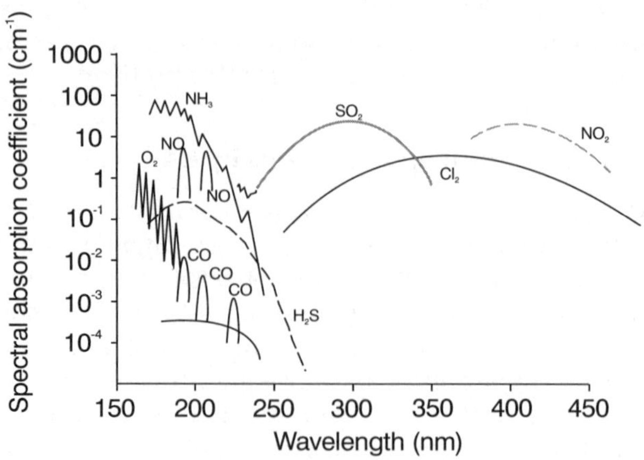

Fig. 8.9 Absorption spectra in the UV range of gases which can be measured by this technique.

the dispersed spectrum. Each diode in the array is monitoring a different part of the spectrum.

Obviously, where two components of the sample have similar absorption characteristics this type of instrument is less able to discriminate between them. The discrete separation of the diodes can limit the resolution of the wavelengths. Figure 8.11 illustrates schematically the layout of the diode array spectrophotometer.

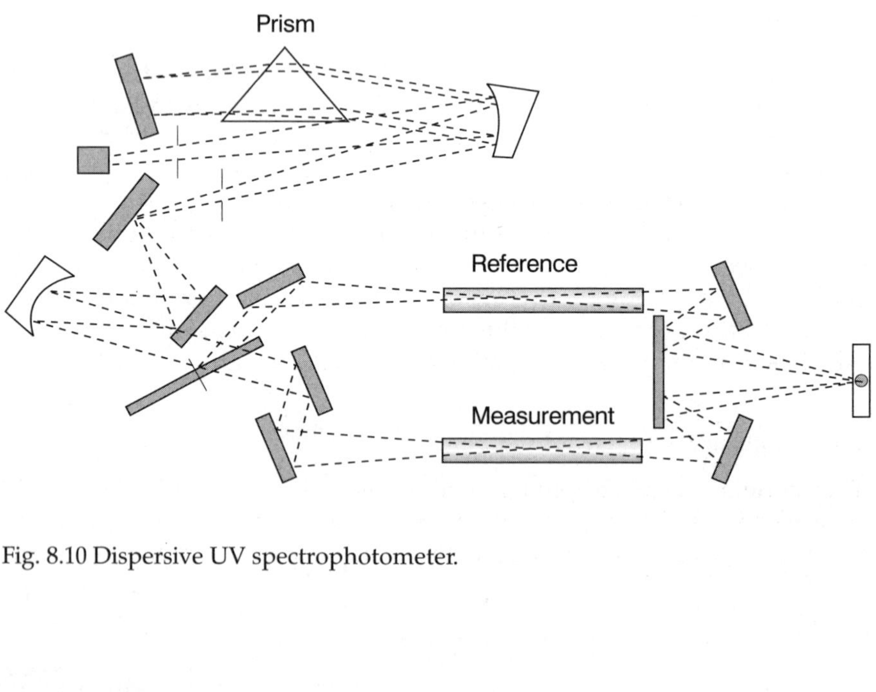

Fig. 8.10 Dispersive UV spectrophotometer.

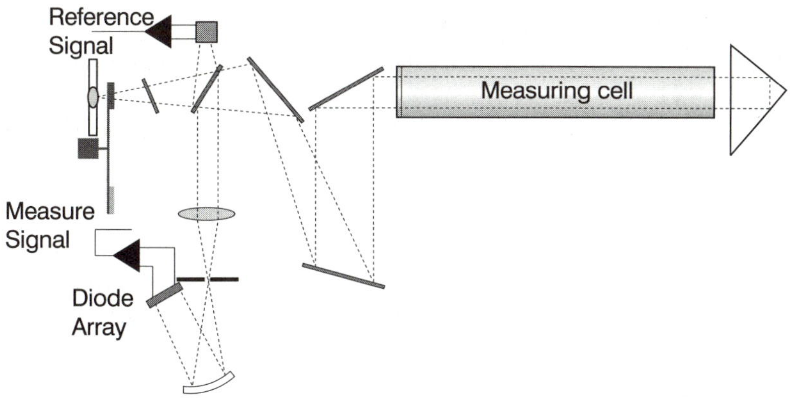

Fig. 8.11 Diode array spectrophotometer

8.4.2 Differential optical absorption

The differential optical absorption technique (DOAS) is a variation on the dispersive IR system. However, in this case there is no need for reference path or reference cells. In this situation, a background spectrum is obtained to characterize the optics and act as a reference when no sample gas is present. During measurement the monochromator scans across the spectrum and the resulting spectrum is compared with that of the reference. This spectrum is a composite of the spectra of the various gases in the sample and the reference spectrum. A computer tries to calculate a best-fit curve by comparing the spectrum measured against combinations of the spectra of each gas stored in its memory. When a good fit is obtained the system reports a confidence factor based on the standard deviation. Unknown gases lead to errors as the system depends on the computer knowing which are to be taken into account.

8.4.3 UV fluorescence analyser

The physical process of UV fluorescence has already been described and turning that into a working analyser is relatively simple. Figure 8.12 illustrates a typical UV fluorescence analyser for measuring SO_2.

In this case a chopper disc produces pulses of UV radiation which pass into the reaction chamber. Electrons in the molecules are pumped up into an excited state and decay back to the ground state. During decay, infrared photons are emitted and these are measured by the detector. The detector is shielded from the lamp by mounting at 90° to the path of the light and uses optical interference filters to reduce sensitivity to emitted and reflected UV.

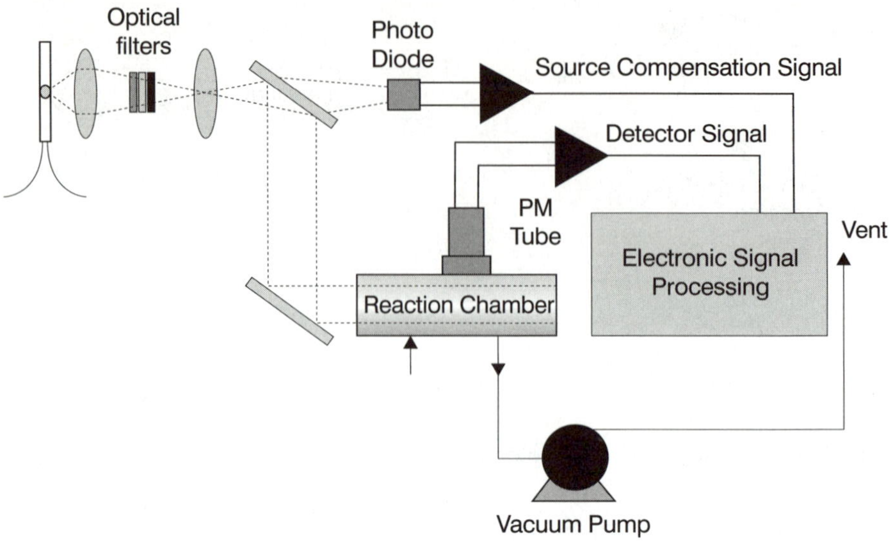

Fig. 8.12 UV fluorescence analyser.

Further optical interference filters narrow the wavelength of radiation emitted by the source to reduce cross-sensitivity caused by other molecules fluorescing. This technique is highly sensitive and is capable of measuring from low parts per million down to low parts per trillion. Unfortunately, it is also highly cross-sensitive to hydrocarbons; consequently, it is mostly used in monitoring SO_2 in ambient air.

Note: Many gases respond to UV by fluorescing and emitting IR, giving a high probability of cross-sensitivity when compared with UV absorption.

8.4.4 Conclusion

UV analysers share many of the same processes – in both the way that they generate UV light and physically react with it to produce a measurement. The technique has many strengths which offer advantages over other methods such as IR. Many of the cross-sensitivities experienced by IR can be avoided by using UV.

8.5 CHEMILUMINESCENT ANALYSERS

8.5.1 Principle

In the chemiluminescent analyser, a chemical reaction creates a molecule with excited electrons which in turn produce detectable radiation during decay to the ground state. In the most common example, ozone is reacted with nitric oxide to form nitrogen dioxide. The radiation emitted occurs in the near-IR and extends from 0.6 to 3.3 microns (Fig. 8.13(a)).

In practice, only the region around 0.6 to 0.9 microns is used. In some instruments the detector window and an optical filter cut off the rest of the spectrum (Fig. 8.13(b)). The phenomenon was first observed by Strutt in 1911 while investigating electrical discharges in air.

Sensitivity increases with decreasing pressure until reaching an optimum level at between 5 and 25 millibars absolute. Further decrease in pressure may cause loss in sensitivity due to insufficient reacting molecules. There is sufficient sensitivity at atmospheric pressure to measure down to low ppm.

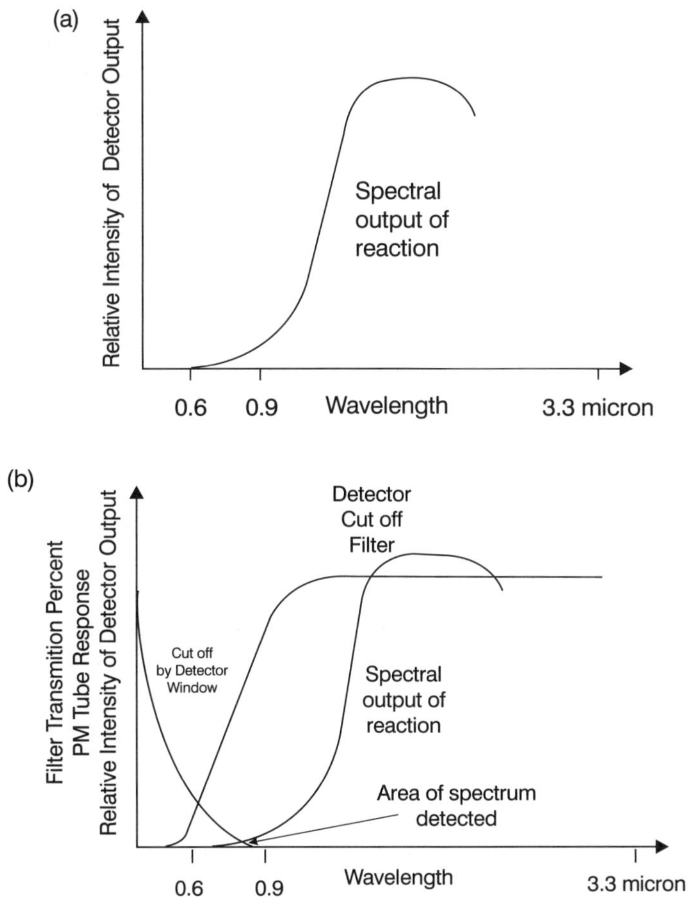

Fig. 8.13 (a) Spectral output of the NO, ozone reaction. (b) Area of spectrum detected.

8.5.2 Quenching

There are several processes by which the excited electrons in the molecules lose energy, and not all excited molecules will emit photons by spontaneous decay to the ground state. Only approximately 7% of all reactions create an excited molecule which decays by the emission of a photon. The processes involved are the same as those previously mentioned in UV absorption.

Quenching is the process where excited electrons do not emit photons but lose some of their energy during collision with other molecules. This occurs when larger molecules such as water vapour and carbon dioxide are present in the sample. At higher pressures, the probability of a collision is greater and therefore quenching is more of a problem. Photons emitted by the reaction may also be re-absorbed.

A typical figure expected for quenching in a vacuum analyser is $<<1\%$ reduction in value for 10% change in CO_2 and $<<1\%$ reduction for 3% change in H_2O. Quenching values in non-vacuum analysers may be considerably higher than this and will be dependent on the effectiveness of the reaction chamber design.

8.5.3 Linearity

Chemiluminescent analysers are linear providing there is excess ozone in the reaction chamber,

i.e. $O_3 >> NO_x$.

Non-vacuum analysers operating at atmospheric pressure will require substantially more ozone; therefore, when high concentrations are to be measured it may be necessary to use pure oxygen to enable the ozonizer to generate sufficient ozone in order to prevent non-linearity.

8.5.4 Cross-interference

Other than quenching effects there are no gases present in normal emission measurements which cross-interfere with the chemiluminescent process. Therefore, a chemiluminescence analyser is highly specific to NO.

8.5.5 Pressure and flow

Most forms of gas analysis respond to pressure in accordance with Boyle's law and Dalton's law, giving them a positive response to barometric pressure, i.e. a 1% increase in barometric pressure produces a 1% increase in detector output. The chemiluminescent analyser is unusual in that sensitivity increases with decreasing pressure.

Flow is a critical parameter. As the detector is measuring the result of a chemical reaction, sample flow into the reaction chamber must therefore be controlled to keep the ratio of ozone and NO_x stable. This is normally controlled by regulators and capillaries, but may also use a critical flow orifice as an alternative.

8.5.6 Typical analyser function

NO_x measurement is the most common application for the chemiluminescent reaction. However, it is only the oxidation of NO to NO_2 which produces any chemiluminescence; therefore, any NO_2 in the sample is not detected unless first converted to NO by a reduction catalyst.

Figure 8.14 shows a schematic of the building blocks in a typical chemiluminescent NO_x analyser.

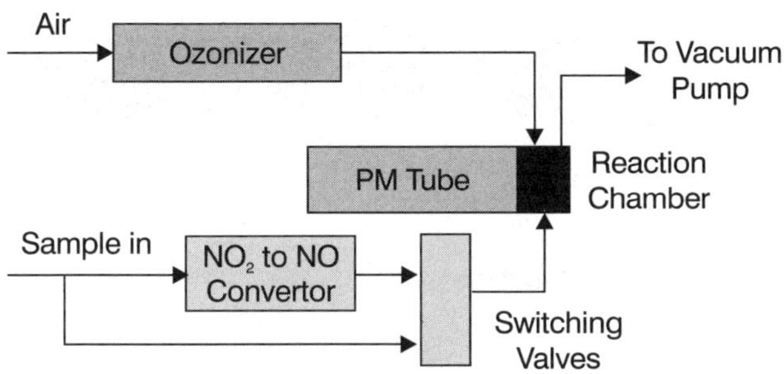

Fig. 8.14 Schematic diagram of a chemiluminescent analyser.

The major items are identified and show the catalytic converter with the switching valve bypass system to produce a value for NO_2 by the subtraction of NO from NO_x.

$$NO_x - NO = NO_2$$

It is common for analysers to have an automatic switching mode enabling circuits rapidly to determine the above calculation and produce separate signals for NO_x, NO and NO_2. Within chemiluminescent analysers there are several variants of instruments based on: (i) three variations of ozonizer; (ii) three variations of capillaries; (iii) three common variations of catalyst; and (iv) vacuum and non-vacuum analysers.

Ozonisers

Figures 8.15–8.17 illustrate the three main variations in ozonizers found in chemiluminescent analysers.

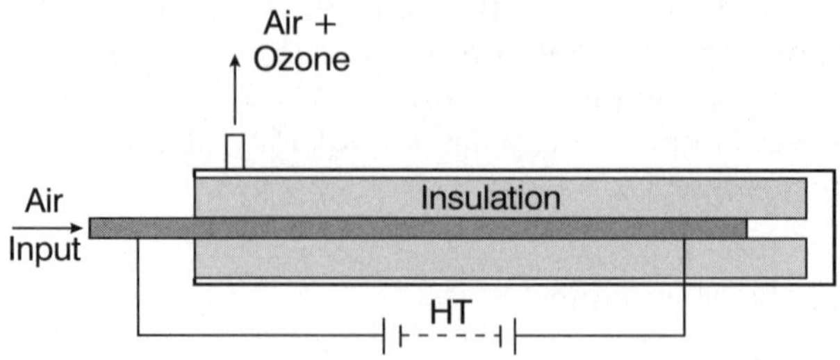

Fig. 8.15 Corona discharge ozonizer.

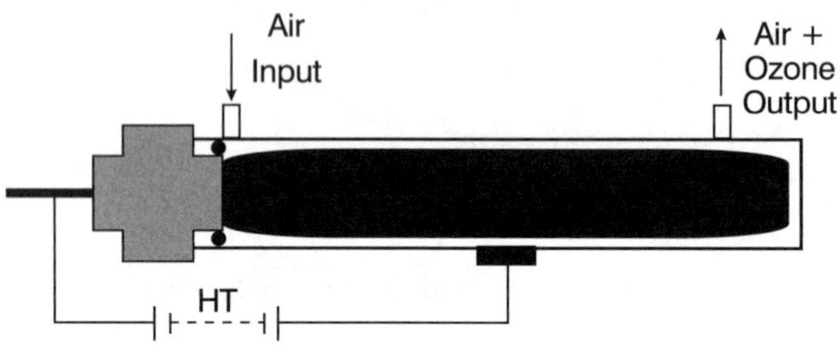

Fig. 8.16 Soft corona discharge ozonizer.

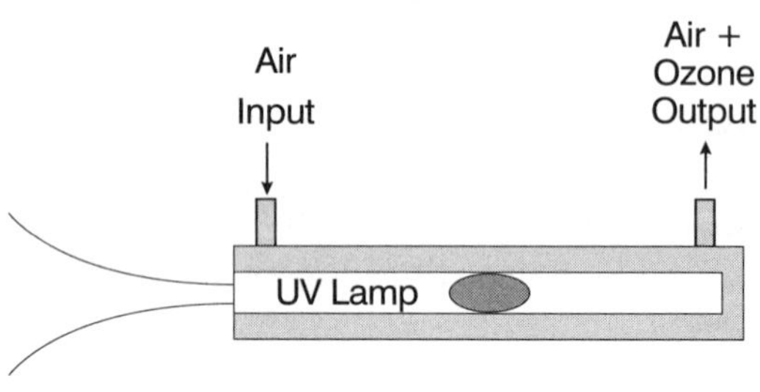

Fig. 8.17 UV ozonizer.

UV lamp ozonizers are similar to the sources used in UV analysers. In this case the air for production of ozone flows over the envelope of the high-energy lamp and ionizes the oxygen molecules, causing some to break down to oxygen radicals. On recombination, a proportion of the free radicals form ozone. This type of lamp may suffer from similar problems in reliability and reduced lifetime as that of the UV sources.

The corona discharge method is the most common and all the early analysers used this type of ozonizer. It is a simple device whereby a glow discharge created by the high voltage across the air and insulation generates a small amount of ozone. Unfortunately, NO also appears and the quantity produced is related to the high voltage. The NO appears as an offset on the analyser zero which is compensated during calibration. At concentrations above 1000 ppm, some analysers must have pure oxygen fed to the ozonizer in order to increase the concentration of ozone and prevent non-linearity.

A soft discharge ozonizer operates on the same principle as that of the corona discharge; however, in this case the central electrode is replaced by a neon lamp. During operation the gas plasma forms the central electrode, providing a large surface area with a very even field-strength across the narrow gap. This allows the ozonizer to run at lower voltages, producing a greater quantity of ozone while avoiding production of NO.

Catalytic converters

The three most commonly used catalytic converters are:

- stainless steel
- molybdenum
- low-temperature carbon

All of these catalysts work well, but it should be noted that none is 100% efficient and all may suffer some form of application-related problems. These can lead to the catalyst being poisoned, which loses conversion efficiency and may not convert any NO_2 to NO.

Stainless steel

With efficiency in the region of 98–99%, this catalyst is fairly reliable for exhaust emission work. However, in reducing atmospheres a stainless steel converter will convert NO to nitrogen. Therefore, manufacturers using this type of catalyst, in that application, normally pre-dilute the sample with air to guarantee an oxygen-rich atmosphere. This may cause some loss of sensitivity.

While stainless steel is normally considered to be non-consumable, at its operating temperature of 600–700°C it can corrode by the production of small pinholes. The high temperature may also stress the components, leading to other problems. Poisoning of this type of catalyst will necessitate complete replacement.

Molybdenum

The efficiency of this catalyst is in the region of 99.9% but, while it is the most efficient, it is rapidly consumed when measuring the higher levels of NO_2 found in exhaust emission gases. This catalyst is normally limited to monitoring NO_x in ambient air.

Low-temperature carbon

The low-temperature carbon catalyst operates at around 400°C or lower. The efficiency of carbon catalysts is slightly lower, in the region of 95–98%; they are also consumable. However, they have the advantages that they are not affected by reducing conditions and are far more resistant to poisoning than stainless steel or molybdenum. As the carbon is gradually oxidized away, it continually exposes new catalyst which is not poisoned. When consumed, a new charge of carbon granules is installed very easily and complete replacement of the converter is not needed. Under normal conditions the lifetime of the catalyst should be approximately 12 months on a vacuum chemiluminescent analyser, depending on its use.

Capillaries

The two most common types of capillary are glass or stainless-steel.

- Glass capillaries tend to be very short with a narrow bore. This makes sample filtration more critical on applications where there is a high particulate present, as blocking of the capillaries leads to analyser failure.
- Stainless steel capillaries tend to be much longer and with a wider bore. Consequently, they are less susceptible to problems caused by dust, particulate or tar deposits from coal, heavy fuel oil, etc.

8.5.7 Conclusion

The chemiluminescent analyser is ideal for measuring NO_x in flue gas and exhaust emissions. The wide dynamic range from ppb to % levels ensures that all concentrations are measured with no cross-interference to other gases. Chemiluminescence is specified by legislation as the technique for certification of gas turbines and internal combustion engines for NO_x.

9

Volatile organics

Stephen Richardson and Nigel Gibson

9.1 BACKGROUND

9.1.1 The pollution issues

Organic compounds contribute to a range of atmospheric pollution problems. The issue that has required the general control of organic compounds is photochemical smog (Seinfeld *et al.*, 1991). Organic compounds also contribute to climate change through global warming and depletion of stratospheric ozone. In addition, some organic compounds are toxic and many are odorous.

9.1.2 Definitions

The term 'organic' means any compound containing carbon and hydrogen. The first studies of the atmospheric chemistry of organic compounds concentrated on hydrocarbons, those which contain only carbon and hydrogen. More recently, interest has widened to include other organic compounds such as chlorinated solvents, alcohols and formaldehyde. The term hydrocarbon does not cover such compounds and so 'volatile organic compound' (VOC) was coined. 'Volatile' is used to distinguish these compounds from the heavier polyaromatic hydrocarbons and dioxins which are normally found in the condensed phase.

VOC is such a broad class of compounds that other terms have been used to describe subgroups of interest: Semi-Volatile Organic Compounds, Reactive Volatile Organic Compounds, Non-Methane Volatile Organic Compounds (NMVOC) to name but a few.

The terms organic compound, hydrocarbon, VOC, SemiVOC, Reactive VOC and NMVOC, etc. are often used interchangeably, causing considerable confusion. This chapter will use the term organic compound unless quoting a source The analytical methods described in this chapter are for organic compounds in the vapour phase. Each method will have certain limitations. For example, they may not be suitable for very light, or very heavy organic compounds. Analytical methods for the heavy organic compounds normally found in the condensed phase, adhering to particulate matter, are dealt with in another chapter.

9.1.3 Principal sources of emission

Organic compounds are ubiquitous. Man-made or anthropogenic sources arise primarily as a result of the petrochemical industry and the use of liquid fuels, organic solvents, plastic, rubber, and other speciality chemicals. Although other sources such as the use of coal, gas, wood and the food industry are important, these are all dwarfed by the emissions associated with the use of oil. The United States Environmental Protection Agency has estimated the following breakdown of reactive volatile organic compound sources (USEPA, 1990):

Source	
Fuel use	*Emission (ktonnes)*
Highway gasoline	4680
Highway diesel	177
Aircraft	190
Railroads	130
Vessels	480
Other off-highway	470
Wood combustion	800
Coal combustion	54
Fuel oil	11
Natural gas	69
Other fuel combustion	7
Industrial processes	
Crude oil production, storage and transfer	540
Petroleum refining	690
Petroleum product, storage and transfer	1530
Organic chemicals	1010
Rubber tyres	50
Plastics	380
Surface coating	2240
Degreasing	370
Dry cleaning	210
Graphic arts	230
Adhesives	40
Solvent extraction processes	40
Textiles	10
Other organic solvent use	280

Food and beverages	170
Iron and steel	170
Other chemicals	570

Reactive VOC emissions relating to the extraction, manufacture and use of petrochemicals amounts to about 90% of all reactive VOC sources.

With such a wide variety of compounds from a bewildering range of sources, the sampling and analysis of organic compounds appears a daunting task. Nevertheless, it is hoped that this chapter will provide an introduction to the various techniques commonly used for capturing, separating and characterizing organic compounds.

9.2 SAMPLING OF ORGANIC COMPOUNDS

9.2.1 Materials

The materials which come into contact with organic compounds during sampling should be inert, of low permeability, smooth and odourless. Common materials used for organic compound sampling include:

- Polytetrafluoroethylene
- Polyvinyl fluoride
- Stainless steel
- Glass

Which materials are used depends on their availability and resistance to heat. Although polytetrafluoroethylene (PTFE) and polyvinyl fluoride are the preferred materials, they are generally not used where temperatures exceed 220°C because of artefact formation and structural breakdown. At higher temperatures stainless steel or glass is used. The type of glass will also affect its resistance to temperature.

9.2.2 Continuous sampling for organic compounds

A typical extractive system, based upon US EPA Method 25, 1983, used for the continuous monitoring of organic compounds in the vapour phase is shown in Fig. 9.1 (USEPA, 1983). Continuously sampling organic compounds directly into an analyser enables a detailed picture of how the emission varies with time to be developed. This approach is useful for processes which are highly variable and where emissions are extracted via relatively few vents or stacks.

Probe and probe positioning

If the gas is well mixed, then a single point can be sampled. The sampling position should be as far downstream from a fan as possible, ensuring

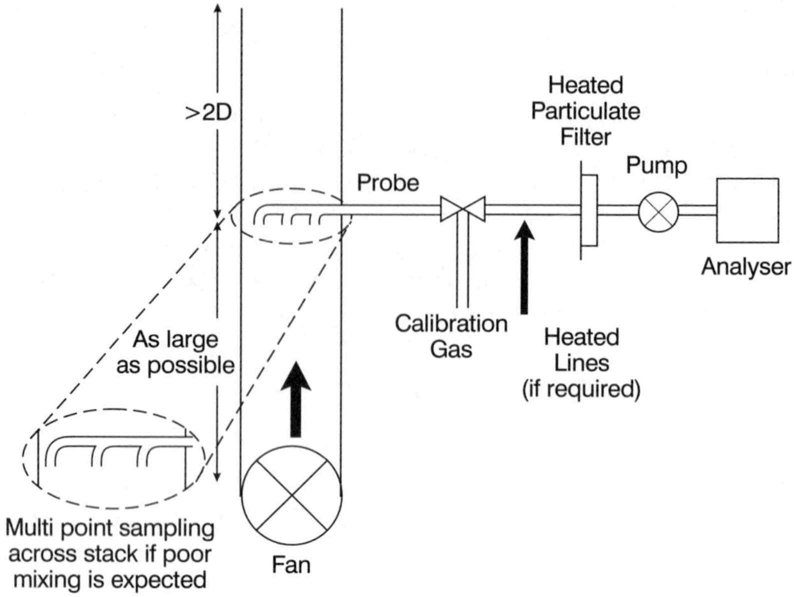

Fig. 9.1. A typical extractive system used for the continuous monitoring of organic compounds in the vapour phase.

good mixing of the gases. In general, the sample probe is positioned at least 1.5 m or two equivalent duct diameters (whichever is less) upstream of the gas discharge to atmosphere.

If the gas is not well mixed, then an uneven distribution of organic compounds across the duct is possible. In this case, the probe should consist of a number of holes to give multipoint sampling. The sample volume extract through each hole should be the same.

Filtration

Almost all analysers require the removal of particulate matter which could block flow control capillaries or foul cell windows. Probes may need to be fitted with filters and most analysers will also incorporate a high-performance filter. The filter medium used must not affect the integrity of the gas sample. The filter may need to be heated to the same temperature as the sample gas to avoid condensation.

Transportation

The sample lines may need to be heated to avoid condensation of organic compounds and other components of the gas stream. If the sample line is

PTFE, then lines are not heated above 220°C. In any case, the sample should not be heated to above the temperature at the sampling point, since this may result in the formation of artefacts and by-products.

Pump

A pump system is used to draw the sample from the stack, through the probe and sample lines. The best position to have the pump is after the analyser. However, this is often not possible because analysers are designed to work under positive rather than negative pressure. The pumping rate is normally as high as possible to minimize the time the sample is in a sample line. The pump may also be heated to avoid condensation.

Gas conditioning

High moisture content and above-ambient stack temperature can lead to condensation within the sample system, resulting in damage to equipment and erroneous results. To prevent this, conditioning of the sample may be required. This can be achieved by using completely heated systems or introducing dry, clean air to dilute the gas, and maintain the gas above its dew-point. Whichever system is used it should be calibrated using standard gases. Permeation tube driers are not generally used for gas conditioning since loss of organic compounds may occur.

9.2.3 Periodic sampling for organic compounds

Periodic sampling may be the only practical method for sampling organic compounds. For example, if a large number of vents or ducts require monitoring, the cost of continuously monitoring all vents may be prohibitive. In some cases, off-site analysis of the organic compound is required.

Unlike continuous monitoring, periodic samples can only give the average concentration of organic compounds over the sampling period. In order to decide whether or not periodic sampling is suitable, two factors have to be considered, the number of samples and the sampling time.

Sufficient samples are required to ensure that the emission is properly quantified, taking account of the likely variation in emissions from a process. For processes which vary greatly, sampling times should be kept as short as possible. In this way, peaks in the emission are not missed. This necessarily means that the number of samples must be increased if a representative profile of the emissions is to be obtained.

There are no hard and fast rules about the number of samples and the sample time required for a given process emission. Nevertheless, there must come a point where the variation of the process emission with time is so great that meaningful results cannot be obtained using a series of periodic samples. In such cases the sampling and analysis should be continuous for the complete emission cycle.

Sampling into bags

Techniques for sampling into bags are usually based on US EPA Method 110, 1980 and one example is shown in Fig. 9.2 (USEPA, 1980). The sample remains in its gaseous form and is taken away for later analysis or concentration/separation before analysis. The sample is analysed or transferred as quickly as possible to prevent sample loss or deterioration; almost always within 24 hours.

A sample bag, either empty or partially filled with clean dry diluent gas and constructed of polyvinylfluoride (although sometimes aluminium bags are used) is sealed within a rigid plastic drum with a volume of approximately 60 litres. By withdrawing air from the space between the bag and the rigid drum, a sample of the gas stream is drawn into the bag. The time over which the sample is collected can be varied by adjusting the rate at which air is pumped from the rigid drum. The apparatus is fitted with a pressure relief system to avoid bursting the bag. Once the bag is filled, the sample can then be transferred, via a concentration/separation step if necessary, to the analyser.

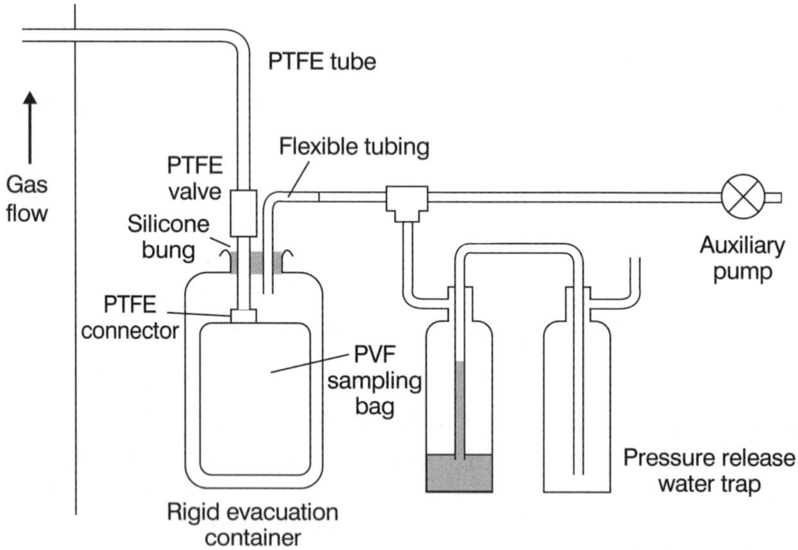

Fig. 9.2 Example of a technique for sampling into bags.

If the gas source is hot, moist or is particularly concentrated, the bag sampling system can be used statically to dilute the gas sample. This pre-dilution is done by measuring the total volume of the bag within the rigid drum using a standard volumetric gas meter. Before taking the sample, the bag is completely filled with clean, dry air or nitrogen. A measured volume of clean air is then removed and the sample taken as described above.

Evacuated steel canisters have been used to sample organic compounds in ambient air. It is therefore possible that canisters may be used to take samples from stacks However, stack concentrations of organic compounds and other pollutants are generally much greater than ambient concentrations. It is not clear as yet how well canisters will maintain the integrity of high concentration samples (Oliver *et al.*, 1986; Gholson *et al.*, 1989).

Pre-concentration for subsequent analysis

Gas adsorption tubes are a convenient method for pre-concentrating, storing and transporting organic samples before analysis.

The adsorbent material in the tube, or similar container, needs to have a high, but reversible, retention for the organic compounds in the sample. The adsorbent must not be susceptible to attack by other pollutants which may be present, for example NO_x, SO_x, HCl, and be capable of storing the bound organic components in a stable form, without losses, until such time as the analysis can be carried out in the laboratory. Losses may be quantified by analysing duplicate samples at different times, to establish the rate of loss of the sample with time.

A variety of adsorbents are available and their suitability depends on the organic compounds to be sampled and the method use to recover the bound material for analysis. For example, some organic compounds adsorb so strongly to activated carbon that thermal desorption alone is insufficient for complete desorption.

If the sample contains organic compounds with very different adsorbing properties, then two different adsorbing materials may be placed in series to catch different sets of organic compounds. The 'stronger' adsorbent is placed after the 'weaker' adsorbent. In this way, the strongest adsorbers are not used to trap the 'stickiest' organic compounds, which would otherwise make subsequent desorption and analysis difficult. Table 9.1 lists the limits of application of common adsorbents.

Volatile Organic Sampling Train

The Volatile Organic Sampling Train or VOST allows organic compounds to be sampled from a gas stream directly onto adsorbent material (USEPA, 1987; Jayanty and Tompkins, 1990). The apparatus is shown in Fig. 9.3.

Table 9.1 Limits of application of common adsorbent materials

Adsorbent	Desorption technique	Limits of application
Tenax	Thermal	Poor retention of compounds with boiling points below 100°C, alcohols and acids
Carbon products	Organic solvents	Poor retention of compounds larger than dodecane
Chromosorb and Poropaks	Thermal	Poor recovery of compounds with high boiling points
Ambersorb products	Thermal	Unsuitable for high boiling point hydrocarbons
Silica gel	Thermal	Aborbs H_2O and CO_2

The VOST is designed to prevent sample break-through on the adsorbent, by cooling the sample gas before it enters the adsorbent traps. Cooling is not appropriate if water or sample condensation is possible before the adsorbent. Two traps are used in series to check if any breakthrough of organic material has occurred. Both tubes are analysed and no organic material should be present in the second tube.

Using bags
As an alternative to sampling directly from the stack, an adsorbent tube is attached to the outlet of a sample bag and a known volume of sample is drawn through it. If the gas concentration is likely to be high, adsorbent cartridges in series should be used to check for breakthrough.

9.3 SEPARATING ORGANIC COMPOUNDS

9.3.1 Gas chromatography

If a mixture of organic compounds is involved and the proportions of the different compounds unknown, then gas chromatography (GC) may be used to separate the compounds (US EPA, Method 18; Mindrup, 1978).

Samples for GC can be highly concentrated organic vapour obtained from an adsorption tube or whole air samples if the compounds are light and unlikely to adsorb onto adsorption tubes. The sample is introduced to one end of a column. The column is packed or coated with an adsorbent. An inert carrier gas drives the plug of organic compounds through the column; however, the plug does not remain intact. The different affinity that each organic compound has for the adsorbent results in a gradual separation of the mixture of compounds. Given a long enough

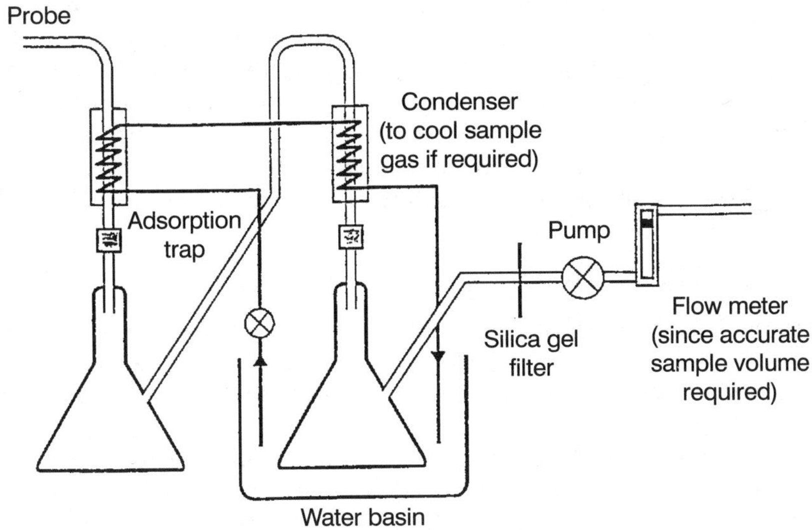

Fig. 9.3 A Volatile Organic Sampling Train.

column and suitable adsorbent, all the organic compounds species can be separated. As a compound leaves the column it is analysed by one of the methods described later. The compound can be identified and quantified by the introduction of known standards.

Even with GC separation, it is not possible to 'run blind'. Some idea of which compounds are present in the sample is still required. Tentative identification can be achieved using column retention times, although these may be influenced by the presence of other compounds. The use of a mass spectrometer as the analyser enables more confident identification of unknown compounds.

9.3.2 Liquid chromatography

Mixtures of organic compounds may be separated by liquid chromatography. The principle is similar to that of GC except that instead of a carrier gas, the organics are in solution with the solvent acting as the carrier.

Classic liquid chromatography consisted of a column packed with a finely divided solid. The solution of organic compounds is allowed to percolate through the column under gravity. Different organic compounds will pass through at different rates, because of their different affinities with the solid phase. The separate components may then be collected at the base of the column.

The principal drawback with this process is that it is rather slow, since compounds move through the column under gravity. However, high-pressure pumps can now be used to speed up the process. Modern High Performance Liquid Chromatography (HPLC) is used when samples are taken into solution, or when derivatives of an organic compound are taken to aid identification. Unlike GC, liquid chromatography is not limited by the volatility or thermal stability of the organic compound.

9.4 ANALYSIS OF ORGANIC COMPOUNDS

Organic compounds have very different chemical and physical properties and the analysis of a mixture of organic compounds is necessarily approximate, unless the different compounds are separated first, e.g. by chromatography.

How approximate the analysis is depends on the analyser and the mixture of organic compounds. For example, if the mixture is of the light hydrocarbons, methane, ethane and propane and the analyser is a flame ionization detector (FID), the relationship between analyser response and mass of organic compounds is close to linear. A number of sources may only have a single organic compound in the emission, or a simple mixture of a constant proportion. In these cases the analyser may be calibrated with the specific compounds of interest and the relationship between analyser response and mass of organic compounds can be established.

For more complicated emissions, for example an animal renderer or surface-coating operations using a range of solvents, any analysis technique which does not identify the separate compounds and their proportion must necessarily be very approximate.

The properties of organic compounds normally used for analysis are:

- absorption of infra-red radiation
- ion production in a flame (FID)
- ion production in a light beam (photo-ionization detection, PID)
- heat evolved during combustion

9.4.1 Infra-red detector

These monitors use a filter to select a particular frequency band of the light which relates to the C–H bond stretch in organic compounds. The simplest form of detector consists of a source, an optical filter, the sample cell and detector. The sample cell may also be used for the calibration gas.

Infra-red detectors are simple to operate and robust. However, they suffer greatly from interference, particularly from water vapour, and the relationship between absorption and compound mass varies considerably from one compound to another (US EPA, 1983; Clayton and Davis, 1989).

9.4.2 Flame ionization detector (FID)

Covalent organic carbon molecules are easily ionized in a flame. An electric field applied across the ionized gas cloud causes a current to flow between the electrodes, depending on the charged species present and the structure of the molecule. This current is approximately proportional to the mass flow of the carbon atoms and generally only varies slightly for a given type of organic compound and carbon number (US EPA, 1983; Clayton and Davis, 1989). The detector is shown in Fig. 9.4.

The combustion chamber contains the hydrogen/air burner and the sample is introduced into the flame together with the combustible gas, either hydrogen or a hydrogen–helium mixture. Gas flows are controlled by calibrated restrictors and all instruments have electronic ignition. The burner itself forms one electrode, with the other suitably positioned to collect the maximum ionization current. With no sample, the hydrogen flame produces only a small current, but with constant flow of sample, or calibrant, a current is generated which can be suitably amplified.

This type of detector has high sensitivity and little or no response to water, carbon dioxide and carbon monoxide. By using catalysts, FIDs are able to distinguish between methane and non-methane hydrocarbons. Unfortunately, the instrument is less sensitive to oxygenated organic compounds such as methanol, formaldehyde and formic acid. In addition, changes in pressure bought about by condensation in the lines can be a major problem, although many FIDs now have internally heated lines. The use of FID generally involves the transport of gas cylinders for fuel use and calibration.

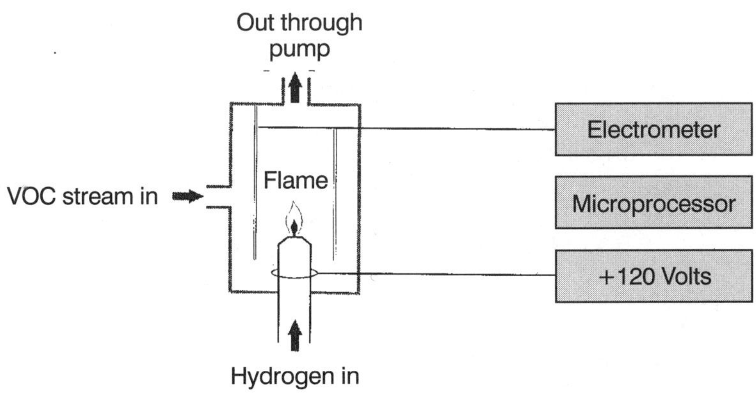

Fig. 9.4 Schematic diagram of a flame ionization detector.

9.4.3 Photo-ionization detection

Organic compounds may also be ionized by photons in a light beam and photo-ionization detection (PID) is similar in principle to flame ionization detection. The advantage is that no fuel is required, so hydrogen and air cylinders can be dispensed with. There is also no zero gas as there is no residual current, unlike the FID. However, correlation between a compound's mass and the detection response is poor and most detectors will not pick up methane (Harrison and Perry, 1986). Photo-ionization detectors are often used as hand-held monitors or as detectors attached to gas chromatographic instrumentation.

9.4.4 Analysis using heats of combustion

When a sample of organic compounds in air is burned, the heat evolved is dependent – to some extent – on the mass concentration of the organic material in the sample. This has been used to develop simple hydrocarbon detectors. A typical detector would use a platinum catalyst to oxidize the sample, the heat evolved would raise the temperature of the catalyst, and this would be monitored by a thermocouple.

Monitors using heats of combustion are robust detectors suitable for field work. These were some of the first detectors to be used on-line to monitor organic emissions and to ensure that lower explosion limits were not exceeded. In general, they are not particularly sensitive and the response may be poor for partially oxidized, nitrated or chlorinated compounds.

9.4.5 Chemical analysis

Another approach is to identify and quantify the organic compounds present by chemical analysis. Chemical analysis may be carried out immediately, or samples taken back to the laboratory for off-site analysis.

Gas detection tubes are used for immediate analysis, although their accuracy is debatable since they are subject to interference by other species.

Off-site chemical analysis is often termed wet chemical analysis as the sample is normally taken into the liquid phase by bubbling through a solution, containing a trapping reagent if necessary. Once trapped the organic compounds may be separated by liquid chromatography if necessary and identified by a variety of methods, suitable for the compound, including:

- coulometry
- titration
- colorimetry

For example, formaldehyde may be trapped in an aqueous solution of metabisulphite, and then treated with concentrated sulphuric acid and chromatropic acid to yield a pink solution which may be analysed colorimetrically.

A number of standard chemical analyses of this type have been produced by the United Kingdom Health and Safety Executive and are described as Methods for the Determination of Hazardous Substances (MDHS).

9.4.6 Long path monitors

Long path monitors were originally designed for air quality assessment both in the ambient atmosphere and from diffuse sources (Harrison and Perry, 1986).

The principle is to direct a beam of light (e.g. visible, infra-red or UV) along a path of typically 100 metres. The light is returned by a retro reflector. Organic compounds present along the path will absorb some radiation, and this absorption can be detected by comparing the reduction in intensity of light in the absorbing frequency, with that of light of a similar, but non-absorbing frequency. Interference is avoided either by very finely tuning to a specific wavelength for a single species, or sending out a pulse of broad band radiation and looking for the characteristic pattern of absorbances from a given species.

As well as noting a single absorption, the fact that organic compounds absorb a range of wavelengths can be used to identify them. The data analysis required to separate the particular fingerprint of absorption from the background absorbances is complicated and this technique can be expensive. The approach works best for organic compounds with particularly unique absorbances.

The advantage of long path monitoring is that it enables emissions from a source to be assessed without recourse to point source monitoring or even the need to enter the plant. As well as assessing air quality, long path monitors can be used to identify leaks along a pipe line and quantify emissions from diffuse sources such as oil refineries, landfill sites or a field of cows.

9.4.7 Future trends

The current trend in industrial monitoring of organic compounds appears to be towards transportable GCs and mass spectrometers for the specialist, with more user-friendly qualitative detectors for everyday industrial use (Kerfoot *et al.*, 1990). As technology improves, reducing costs, long path monitors will become more widely available

A great deal of effort has been placed on improving the accuracy of analysers. Now, the major errors associated with the monitoring of organic compounds occur during sampling and field analysis rather than analysis at the laboratory. Given the importance of monitoring emissions for the demonstration of compliance with regulations as well as process control, considerable effort must now be placed on the development of highly trained sampling teams using portable, robust and validated equipment.

REFERENCES

Clayton P. and Davis B. (1989) *HMIP Manual on Environmental Sampling and Analysis. Section B - Gaseous Pollutants*, Warren Spring Laboratory, UK, LR680 (PA). ISBN 0 85624 540 2.

Gholson, A.R., Storm, J.F., Jayanty, R.K.M. *et al.* (1989) Evaluation of canisters for measuring emissions of volatile organic air pollutants from hazardous waste incineration. *Journal of the Air Pollution Control Association*, **39**, 1210–17.

Harrison, R.M. and Perry, R. (1986) *Handbook of Air Pollution Analysis*, Chapman & Hall, London. ISBN 0 412 24410 1.

Health and Safety Executive Publications Catalogue, HSE Books, PO Box 1999, Sudbury, Suffolk, UK.

Jayanty, R.K.M. and Tompkins, S.B. (1990) Summary of POHC audit results using VOST and bag methods during RCRA trial burns. *Journal of Air Waste Management Association*, **40**, 1505–9.

Kerfoot, H.B., Pierett, S.L., Amick, E.N. *et al.* (1990) Analytical performance of four portable gas chromatographs under field trial conditions. *Journal of Air Waste Management Association*, **40**, 1106–13.

Mindrup, R. (1978) The analysis of gases and light hydrocarbons by gas chromatography. *Journal of Chromatographic Science*, **16**, 380–9.

Oliver, K.D., Pleil, J.D. and McClenny, W.A. (1986) Sample integrity of trace level volatile organic compounds in ambient air stored in Summa polished canisters. *Atmospheric Environment*, **20**, 1403–11.

Pleil, J.D. et al. (1985) *Storage Stability of Volatile Organic Compounds in Summa Polished Canisters*, Presentation to the 78th Annual Meeting of the Air Pollution Control Association, June 16–21.

Seinfeld, J.H. (ed.) (1991) *Rethinking the Ozone Problem in Urban and Regional Air Pollution*, National Academy Press, Washington, DC. ISBN 0-309-04631-9.

US EPA (1980) *Method 110. Determination of Benzene from Stationary Sources*, EPA Stationary Source Sampling Methods (NSPS) and NESHAPS). McIlvaine Company, 2970 Maria Avenue, Northbrook, Illinois, USA.

US EPA (1983) *Method 25. Determination of Total Gaseous Non Methane Organic Emissions using a Flame Ionisation Detector*, EPA Stationary Source Sampling Methods (NSPS) and NESHAPS). McIlvaine Company, 2970 Maria Avenue, Northbrook, Illinois, USA.

US EPA (1987) *Municipal Waste Combustion Study – Sampling and Analysis of Municipal Waste Combustors*, United States Environmental Protection Agency Report EPA/530-SW-87-021f, Washington, DC.

US EPA (1988) *Method 18: Measurement of Gaseous Organic Emissions by Gas Chromatography*, EPA Stationary Source Sampling Methods (NSPS) and NESHAPS). McIlvaine Company, 2970 Maria Avenue, Northbrook, Illinois, USA.

US EPA (1990) *National Air Pollutant Emission Estimates 1940–1988*, United States Environmental Protection Agency Report EPA-450/4-90001, Research Triangle Park.

10

Sampling and analysis of PAH, dioxins and furans

Paul T. Williams

10.1 INTRODUCTION

Of the organic micropollutants emitted from combustion systems, the chemical groups comprising, dioxins and furans, and polycyclic aromatic hydrocarbons (PAH) are of most concern because of their associated health hazard. The concern over dioxins and furans arises from a number of animal studies which show that for some species they are extremely toxic at very low levels of exposure. The extrapolation of these animal data to humans, though contentious, has led to dioxins and furans acquiring their notoriety as one of the most toxic group of chemicals known to humans. PAH are also seen as highly toxic since, among the environmental chemical groups, they comprise the largest group of carcinogens. PAH adsorbed on airborne particulates are believed to be major contributors to the higher death rate from lung cancer in urban areas as compared with rural areas.

Polychlorinated dibenzo-*p*-dioxins (PCDD) or 'dioxins' and the closely related polychlorinated dibenzofurans (PCDF) or 'furans' constitute a group of chemicals that have been demonstrated to occur ubiquitously in the environment. They have been detected in soils and sediments, rivers and lakes, chemical formulations and wastes, herbicides, hazardous waste site samples, landfill sludges and leachates. More contentiously, PCDD and PCDF are found in the products of combustion, including the ash, stack emissions, water and other process fluids, from the combustion of municipal waste, coal, wood and industrial waste.

PAH also have a wide distribution in the environment, for example occurring in sediments, fossil fuels and in combustion products from natural combustion in forest fires. However, the major source of PAH emissions are from anthropogenic sources, which include for example, oil- and coal-fired power generating plant, coke production, municipal waste incineration, residential furnaces and from diesel and gasoline engines.

The determination of PCDD and PCDF, and PAH emissions from combustion systems present difficult sampling problems because of their very low concentrations and their perceived toxicity. Analytical problems stem also from their low concentration and their separation from the other organic species present in the sample which may be in high concentration and the separation and identification of the large number of individual chemical species.

10.2 CHEMISTRY AND TOXICOLOGY

The generalized molecular structures of PCDD and PCDF are shown in Fig. 10.1, tricyclic aromatic compounds containing two (dioxin) or one (furan) oxygen atoms. Each of these structures represents a whole series of discrete compounds having between one and eight chlorine atoms attached to the ring substituting for hydrogen. For example, Fig. 10.1 shows the tetra isomers with four chlorine atoms in the 2,3,7,8 positions, i.e. 2,3,7,8-tetrachloro dibenzo-*p*-dioxin (2,3,7,8-TCDD) and 2,3,7,8-tetrachloro dibenzofuran (2,3,7,8-TCDF). Since each chlorine atom can occupy any of the eight available ring positions it can be calculated that there are 75 PCDD isomers and 135 PCDF isomers. All the PCDD and PCDF are solids with high melting and boiling points and limited solubility in water.

PAH are compounds based on aromatic benzene rings which are fused to form two or more polycyclic rings. The chemical class of polycyclic aromatic compounds may be subdivided into different subgroups depending on the heteroatom in the polycyclic aromatic structure. Of interest in combustion streams are, nitrogen-containing PAH (PANH), sulphur-containing PAH (PASH) and nitro-containing PAH (NPAH). Examples of

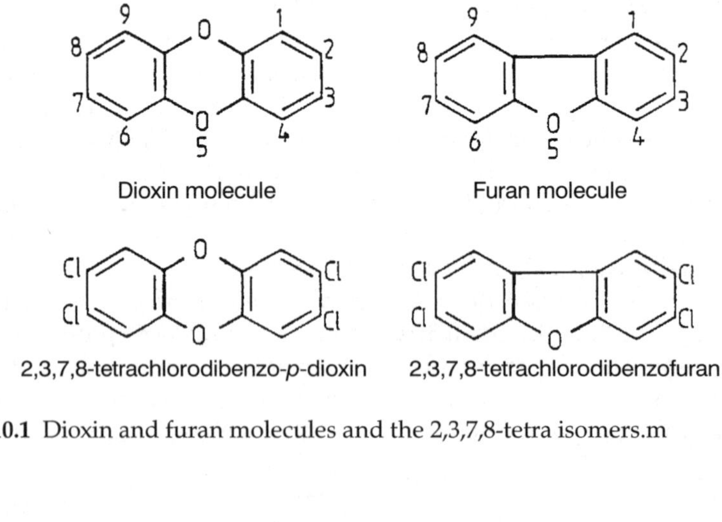

Fig. 10.1 Dioxin and furan molecules and the 2,3,7,8-tetra isomers.m

the different subgroups are shown in Fig. 10.2. Virtually all PAH are solid at ambient temperature, and exhibit low volatility. They are also sparingly soluble in water; however, the presence of other solvents in polluted water can markedly increase the solubilities of PAH.

A comprehensive review (Choudhary *et al.*, 1983) has been published dealing with the health effects of PCDD and PCDF, by far the majority of work being on animal tests. Toxic response has been shown to be a function of each particular animal system and a variety of toxic responses have been recorded depending on animal species. However, it is clear that this group of compounds is extremely toxic. Toxicity depends on the number and position of the chlorine substituents with 2,3,7,8-TCDD being the most toxic. There is less information with respect to the toxic effects on humans and most existing data have been derived from occupational exposure or industrial accident victims. In such cases the effects of exposure to PCDD and PCDF are difficult to interpret since there have been problems in controlled methodologies, inadequate information on intake and exposure mode and level. In addition, exposures have often been to mixtures of PCDDs and PCDFs and also in conjunction with other related and possibly hazardous compounds and a number of workers have suggested that the effects of PCDD and PCDF on humans are inconclusive and require further study.

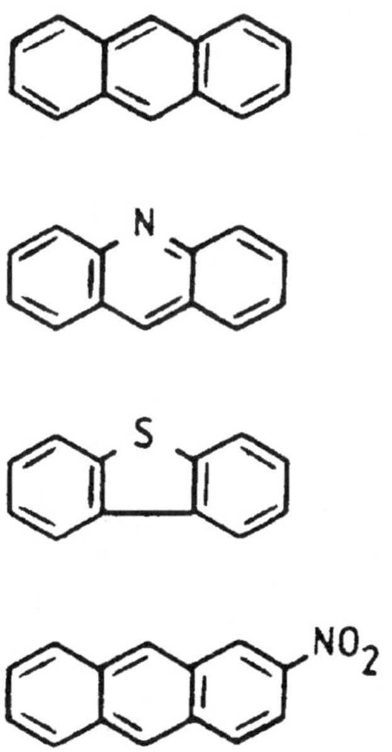

Anthracene PAH

Acridene PANH

Dibenzothiophene PASH

2-Nitroanthracene NPAH

Fig. 10.2 Polycyclic aromatic compounds (PAC).

The potential threat of PCDD and PCDF to humans should be assessed bearing in mind that the 75 PCDD isomers and 135 PCDF isomers have differing toxicities and are often present in multiple mixtures of the isomers. The assessment of the toxicity of PCDD and PCDF mixtures has led to the development of the Toxic Equivalent (TEQ) scheme. This uses the available toxicological and biological data to generate a set of weighting factors, each of which expresses the toxicity of a particular PCDD or PCDF in terms of an equivalent amount of the most toxic and most analysed PCDD – 2,3,7,8-TCDD. Thus, in the most widely accepted method, 2,3,7,8-TCDD would have a TEQ of 1.0 and OCDD for example would have a TEQ of 0.001 (Table 10.1).

Many reviews exist on the health hazard associated with PAH and the relative carcinogenic activity of some PAH have been assessed, for example (Lee *et al.* 1981). Cancers of the human lung, stomach, kidneys, scrotum and liver have been associated with exposure to PAH. However, not all of the very large number of PAH known to exist are biologically active and many have not been tested, either individually or as they occur in complex mixtures. Most work has been undertaken in animal studies with relatively high doses of PAH. It has been found that carcinogenic response to PAH are species-dependent and that PAH exhibit differences in target tissue sensitivity such that for example, certain PAH are significantly more potent carcinogens in lung tissue than on skin. The carcinogenic activity of a particular PAH is dependent on various structural features of the molecule such as shape, size and steric factors and minor structural differences in PAH may result in drastic changes in carcinogenicity. Consequently, isomeric PAH may differ markedly in their carcinogenic activities, for example benzo[a]pyrene has significantly greater

Table 10.1 Toxic Equivalents (TEQ) based on the NATO Committee on Challenges to Modern Society Scheme

PCDD		PCDF	
2,3,7,8–TCDD	1.0	2,3,7,8–TCDF	0.1
1,2,4,7,8–HxCDD	0.5	2,3,4,7,8–TCDF	0.5
		1,2,3,7,8–PeCDF	0.05
1,2,3,6,7,8–HxCDD 1,2,3,6,7,8–HxCDD 1,2,3,7,8,9–HxCDD	0.1	1,2,3,4,7,8–HXCDF 1,2,3,7,8,9–HxCDF 1,2,3,6,7,8–HxCDF 2,3,4,6,7,8–HXCDF	0.1
1,2,3,4,6,7,8–HpCDD	0.01	1,2,3,4,6,7,8–HpCDF 1,2,3,4,7,8,9–HpCDF	0.01
OCDD	0.001	OCDF	0.001

activity than benzo[e]pyrene. Carcinogenicity of PAH has mainly been observed for three-, four-, five- and six-ring compounds.

10.3 EMISSIONS FROM COMBUSTION SOURCES

Organic micropollutants are found in the emissions from a wide variety of combustion systems burning a wide range of materials. The major sources of emission of PCDD and PCDF to the environment from combustion sources are generally regarded as being from municipal waste incinerators while other sources include industrial and domestic coal fires, vehicle exhausts, hospital incinerators and coal-fired power stations. The emissions of PCDD and PCDF from municipal waste incinerators are dependent on the efficiency of the combustion unit and the extent of gas clean-up systems installed. The older type of incinerator with poor combustion efficiency and limited flue gas clean-up can have PCDD and PCDF emission levels of the order of 10^2 to 10^3 ng/m^3. However, the modern incinerator with efficient combustion and extensive gas clean-up systems, will have very low levels of PCDD and PCDF emissions. Indeed, many countries have legislated limits for PCDD and PCDF emissions from incinerators below 1 ng/m^3. The data on emission levels from other combustion sources are limited. The emission of PCDD and PCDF reported from coal-fired power stations have shown zero or trace levels of PCDD and PCDF, while industrial coal burning from limited tests in the UK have shown between 14 ng and 162 ng total TCDF per kg of coal combusted. PCDD and PCDF are produced at low levels from certain motor vehicle exhausts using leaded gasoline, the scavenger dichloroethane acting as a precursor for the PCDD and PCDF. PCDDs and PCDFs have also been detected in used lubricating oil derived from chlorinated additives in the oil or gasoline and have also been detected in the emissions from woodburning stoves and cigarettes.

PAH have a similar association with the incomplete combustion of a wide variety of materials. For example, PAH have been found in the emissions from the combustion of wood, coal, oil, natural gas, municipal and industrial waste and tobacco. The emission of PAH from incinerators firing municipal solid waste and refuse-derived fuel are mostly less than 10 μg/m^3. Benzo[a]pyrene had a flue gas emission concentration of 0.04 μg/m^3. PAH emissions from coal- and oil-fired power stations produced similar concentration levels of PAH as for waste incinerator emissions; for example, benzo[a]pyrene had a reported concentration of 0.05 μg/m^3. PAH have been recorded in the exhaust emissions from gasoline and diesel engines, diesel engine exhaust in particular containing higher levels of PAH.

10.4 SAMPLING

The extremely low concentrations of PCDD, PCDF and PAH in combustion system emissions are clearly a test of the sampling procedures and analytical skills of the investigator, particularly since the emissions are associated with other organic pollutants which may be in much higher concentrations. The formation mechanism of PCDD and PCDF has implications for the sampling procedure and sampling location in the combustion unit. A number of theories have been proposed for the formation of PCDD and PCDF during combustion. For example, it has been suggested that PCDD and PCDF occur as trace constituents in the material input and because of their thermal stability's they survive the combustion process. This route has been suggested in particular for municipal and industrial waste combustion. Alternatively, PCDD and PCDF may be produced during the incineration process from precursors such as chlorinated hydrocarbons. Of direct consequence for the sampling protocol, is the theory that PCDD and PCDF are produced as a result of *de novo* synthesis; elementary reactions of the appropriate elements, carbon, hydrogen, oxygen and chlorine atoms producing PCDD and PCDF. PCDD and PCDF have been shown to form on fly ash containing residual carbon collected within the heat exchanger or electrostatic precipitator of the combustion system at temperatures in the region of 300°C to 400°C in the presence of flue gases containing HCl, O_2 and H_2O. It is thought the reaction is catalysed by various metals, metal oxides and silicates present in the fly ash. The mode of formation of PCDD and PCDF will clearly have consequences for the sampling protocol. If the *de novo* synthesis mode of formation is dominant, then a sampling point which measures PCDD and PCDF before they have formed by the *de novo* route will give a false impression of the emitted concentration.

A further consideration which applies to PAH, PCDD and PCDF sampling protocols is the sampling temperature of the filter and gas phase trapping system. PAH occur throughout the boiling point range; for example, naphthalene with two aromatic rings has a boiling point of 218°C and coronene with seven aromatic rings has a boiling point of 525°C. PCDD and PCDF have a higher boiling point range than most PAH – 2,3,7,8-TCDD has a boiling point of 421°C. The vapour pressure of the PAH also means that even at ambient temperatures significant concentrations of PAH with boiling points up to 400°C will be present in the gas phase. Consequently, the distribution of PAH between the condensed and gas phase will be a factor of the sampling temperature. The distribution between filter trapping of the condensed organic material and cold or resin trapping of the vapour phase material will be dependent on the sample temperature of the filter. Table 10.2 shows the distribution between filter and organic polymeric resin trap for coke oven emissions collected at 50°C and petrol engine exhaust emissions. Even at the relatively low sample temperature the majority of the PAH are present in the gas phase, including the four-ring PAH which are collected in significant concentrations. The vapour pressure of PCDD and PCDF are low compared with PAH and therefore tend to be less volatile; however, PCDD and PCDF will also have a low but significant fraction of the sample in the gas phase and the distribution between filter and gas phase trapping system will be a consequence of the sampling temperature. Hagenmaier *et al.* (1988) sampled from an incinerator stack using parallel systems with one system having the filter at 90°C and the second having the filter at 140°C. A significant difference in the distribution between filter and condensate/vapour adsorption between the two set-ups occurred while the total overall PCDD and PCDF were identical.

PAH, PCDD and PCDF are associated with particles that occur throughout the size range and the smaller respirable size range below about 0.5 micron have been shown to contain increased concentrations of PAH, PCDD and PCDF.

The sampling of all organic micropollutants from combustion systems will be essentially the same. The sample is withdrawn from the combustion system via a probe, cooled to a known temperature and the condensed organic compounds which may be associated with particulate matter are trapped on a glass-fibre or quartz-fibre filter. The gases are then further cooled and passed through a resin trap which traps the gas phase hydrocarbons.

The withdrawal of a sample along the length of a probe and through a filter may cause chemical interactions to take place, with the possibility of forming false particles or artefacts as a result of the sampling process. Interaction of the components of the combustion effluent stream which are at elevated temperatures, and contain complex hydrocarbons including PAH, PCDD and PCDF, water vapour, HCl and sulphur and nitrogen oxides may occur in the sampling system or on the filter. For example, the reaction between PAH deposited on glass-fibre filters with nitrogen

Table 10.2 Distribution of PAH between the condensed and vapour phase (relative %)

PAH	Coke oven emission		Petrol exhaust emission	
	Condensed	Vapour	Condensed	Vapour
Naphthalene	0	100	–	–
Anthracene	19	81	21	79
Fluoranthene	69	31	53	47
Pyrene	60	40	47	53
Chrysene	98	2	–	–
Benzopyrenes	100	0	99	1

dioxide and water vapour has been shown to produce nitro-PAH. Similarly, the *de novo* formation reaction may take place in the sampling system itself, forming artefact PCDD and PCDF.

A number of different sampling systems have been used for the determination of PAH, PCDD and PCDF; however, the majority are based on the US Environmental Protection Agency (EPA) Method 5 for particulates which has been adapted to trap gaseous organic micropollutants, and is the proposed EPA Method 23 (Fig. 10.3). The probe is lined with borosilicate glass or quartz and is temperature controlled and equipped with a pitot tube and thermocouple to measure stack gas flow rate and temperature. Isokinetic sampling traps the coarse and fine particulate on a glass-fibre filter held in an oven maintained at 180°C before the gases are cooled in a condenser and the organic components remaining in the gas phase are removed using an adsorption resin such as XAD-2. The particulate and gaseous traps are followed by a gas clean-up system to remove further acidic vapour and water vapour before the gas-metering system. The system condenses out the flue gases in the presence of XAD-2 sorbent resin and any sulphur oxides, hydrogen chloride and oxides of nitrogen may also condense out and can attack the sorbent media. This attack does not appear to affect the yield of organic micropollutants but may nevertheless produce artefacts which are not present in the original flue gases and prevent further analysis for other components.

After sampling, the entire sampling train or the individual components should be sealed with hexane-cleaned foil, ground glass caps or PTFE tape and stored under refrigerated conditions before analysis.

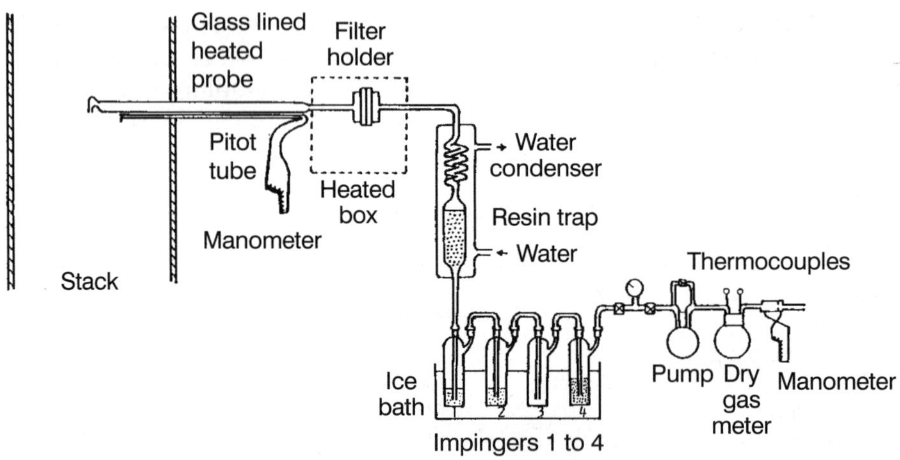

Fig. 10.3 EPA modified Method 5 sampling system for organic micropollutants. (Impinger 1 and 2 = water, 3 = dry, 4 = silica gel).

Quartz- or glass-fibre filters are preferred since they are suitable for high-temperature sampling and may be analysed by solvent extraction without organic interference. The filters should have collection efficiencies of at least 99% for particles of 0.3 micron or larger. The polymeric resins most used have been Tenax and XAD resins and polyurethane foam. Considerations in the use of the resin should include the capacity and breakthrough volume, the possibility of contamination from the resin, and the compatibility of the resin with eluting solvents or thermal stability during thermal desorption.

The use of standards added to the filter before sampling are recommended to determine any sampling errors, for example due to loss of sample or artefact formation. It is recommended that known amounts of at least two internal standards labelled with ^{13}C should be added to the filter before sampling. A further different standard is recommended to be added to the filter after sampling for determination of the analytical recovery. Work has suggested that the internal standards spiked on the filter appear to vaporize to a greater extent than the native compounds in the sample, possibly due to the longer residence time of the internal standards in the hot zone and by a stronger adsorption of the native compounds to the particulates. Variable and low recoveries of some PCDD standard samples from incinerator sampling have been reported, in some cases as low as 5%, with mean recoveries of 51%, 39% and 40% for ^{13}C-labelled hexa-CDD, tetra-CDD and penta-CDD respectively. The losses represented volatilization of the standard since it was found that there was no chemical degradation or rearrangement of the internal standard as the added ^{13}C-labelled PCDD was the only ^{13}C-labelled compound detected.

Other sampling systems have been used for the trapping of organic micropollutants. The system recommended for the Nordic countries consists of a glass-fibre filter followed by a condenser and XAD-2 adsorbent trap. The method differs from the modified EPA Method 5 system in that the condenser traps out the acidic condensate before passing the vapour to the adsorption resin. Thus, the possibility of acid attack on the resin is reduced. However, any soluble organic material may be liable to acid attack in the condensed solution which may contain strong acids. Another sampling system for organic micropollutants used by some workers is based on air dilution of the flue gas sample, the diluting air having been cooled, dried and filtered. During cooling and dilution volatile components such as acid gases and water remain in the gas phase, while less-volatile components including the organic micropollutants condense and are collected on a glass-fibre filter. Since the filter is held at low temperature, further evaporative loss is relatively low and, because there is no acid condensation, there is no artefact formation due to acid attack. A resin adsorption trap may be incorporated after the filter to trap more volatile components and increase collection efficiency.

Because of the very low concentrations of organic micropollutants found in combustion system flue gases, very long sampling times are required to trap sufficient sample for analysis. For example, recommended minimum sampling times for PCDD and PCDF have been variously recommended as 3 hours to 11 hours.

10.5 ANALYSIS

The organic micropollutants may be present in condensed form or associated with particulate material on the filter or as ash samples, adsorbed in the resin representing the gas phase organic compounds or absorbed in condensed water. While high analytical recoveries of PAH, PCDD and PCDF have normally been reported, nonetheless it is recommended that internal standards are added to the sample before analysis to verify the recoveries of the organic micropollutants, since in some cases analytical recoveries have fallen to only 40%.

Analysis is performed in three stages:

1. **Extraction**: extraction of the organic micropollutants from the sample. These may vary from ash, particulate adsorbed on glass-fibre filter paper, condensed water, organic solvent from the sampling train rinse after sampling, and adsorption resin, each requiring a different extraction procedure.
2. **Clean-up**: a series of steps to eliminate compounds which interfere with the analysis and to remove the PAH or PCDD/PCCD as a separate fraction from the extracted sample which may contain many other chemical groups in high concentration.
3. **Final analysis**: the fraction containing the PAH, or PCDD/PCDF are most frequently analysed by capillary gas chromatography to separate the individual compounds and isomers, followed by detection using mass spectrometry.

While the analytical strategy is similar for each laboratory, the solvents used, the liquid chromatography columns and solvents used for clean-up, the columns used for gas chromatography and the final detection method may vary.

10.5.1 Extraction (separation)

Filter paper and ash extraction

Filter paper and ash extraction is usually carried out by Soxhlet extraction and is recommended in some standard methods. The technique involves continuously contacting the collected material with boiling organic solvent for a number of hours, the solvent being automatically recycled by a siphoning system. Various solvents have been recommended for the

Table 10.3 Extraction efficiencies for organic micropollutants using various solvents

(a) Extraction efficiency of ^{14}C – benzo[a]pyrene from diesel exhaust particulates

Solvent	Relative efficiency (%)
Toluene	100
Xylene	98
Benzene	98
Dichloromethane	87
Cyclohexane	58
Acetone	36

(b) Extraction efficiency of PCDD and PCDF from incinerator fly ash

Solvent	Relative efficiency (%)	
	PCDD	PCDF
Toluene, acid-treated Soxhlet extraction	100	100
CH_2Cl_2 Soxhlet extraction	6	11
CH_2Cl_3 Soxhlet extraction	13	20
Acetone/hexane ultrasonic agitation	1	1
CHCl manual shaking	1	2
Toluene extraction of a fly ash solution in HF	9	16

extraction of organic micropollutants for example, dichloromethane, benzene and toluene) Some solvents may be more suitable for a particular matrix – acetone has been reported as nearly 100% efficient for the extraction of PAH from air particulates, but is less efficient for the extraction from diesel particulates. Table 10.3 shows the relative extraction efficiencies of various solvents for the extraction of radioactively labelled benzo[a]pyrene from diesel exhaust particulates and for PCDD and PCDF from municipal solid waste fly ash (Lustenhouwer *et al.*, 1980; Scheuntzle and Perez, 1981). Acid pre-treatment of the fly ash has been used in some cases to increase extraction efficiency for PCDD and PCDF before Soxhlet extraction. The use of acid is to attack the surface of the fly ash to expose the inner surfaces which may contain PCDD and PCDF that are not simply adsorbed on the surface. However, acid pre-treatment – particularly the use of hydrochloric acid – may result in artefact formation altering the original composition of the organic components. Table 10.3 shows that, compared with Soxhlet extraction using toluene following acid pre-treatment, very low recoveries of PCDD and PCDF were found using other methods.

Mixtures of solvents have been found to be more effective for PAH extraction from certain sample types. For example, benzene : methanol (4 : 1) has been found to extract PAH efficiently from diesel particulates, the presence of methanol leading to increased extraction of heterocyclic PAH (Williams *et al.*, 1986). Extraction of the organic compounds is essentially complete after 8 hours and normal extraction times are between 6 and 8 hours. Extraction of PAH may be difficult from materials on which they are strongly adsorbed such as carbon black and fly ash; for example, extraction efficiencies of less than 30% for benzo[a]pyrene have been found for fly ash.

Since the organic micropollutants are usually present in trace amounts in combustion effluents, care must be taken to avoid the introduction of contaminants during extraction procedures, particularly since the analysis usually involves concentration steps. Therefore, solvents should be distilled in glass and checked for purity before use, and cellulose Soxhlet thimbles should be pre-extracted. The apparatus should also be wrapped in aluminium foil which minimizes photodegradation of the organic compounds and also acts as insulation to maintain a constant temperature profile in the extraction apparatus. Normal analytical procedures of using clean glassware and Teflon tubes for solvent evaporation should be used.

Ultrasonic agitation using high-intensity ultrasonic vibration (20 kHz) with an organic solvent at room temperature is an alternative to Soxhlet extraction. Increased extraction of PAH from air particulate samples using cyclohexane with ultrasonic agitation compared with Soxhlet extraction have been reported, but the method is less efficient for some samples, such as diesel exhaust particulate PAH. The advantage of ultrasonic agitation is the much shorter extraction time, of about 30 minutes, compared with 8 hours required for Soxhlet extraction.

Condensed water extraction

The condensation of water vapour in the combustion system or sampling system may result in condensed organic compounds in the water. The organic compounds are extracted by liquid–liquid partition, in this case, water condensate–immiscible organic solvent partition. Typical solvents for PAH extraction are dichloromethane, benzene and toluene, while for PCDD and PCDF, benzene, toluene and hexane have been used. The solvent is added as an immiscible layer to the water condensate and shaken or stirred to give intimate mixing of the liquids. The organic compounds preferentially partition or extract into the organic solvent. Multiple partitioning increases the extraction efficiency.

Polymeric organic resin extraction

Vapour phase organic micropollutants adsorbed on polymeric organic resins are recommended to be extracted by solvent Soxhlet extraction. The solvent should be chosen such that it does not react with the polymeric resin. Pentane has been used for PAH extraction from Tenax resin) methylene chloride from XAD-2 resin, and benzene and toluene have been used for the extraction of PCDD and PCDF from XAD-2 resin.

10.5.2 Sample clean-up

The extract from the filter, ash, water condensate and polymeric resin trap contain not only PAH and PCDD/PCDF but also many other classes of organic compound such as, alkanes, PCBs, phthalate esters, phenols, etc. These compounds may be in much higher concentrations than the PAH or PCDD/PCDF and therefore a clean-up stage is required to remove these interfering species and isolate a PAH- or PCDD/PCDF-rich fraction. The extracts from the filter, ash, water condensate and polymeric resin trap may be analysed separately or together to give the bulk emission. The extract is concentrated by the vaporization of the extracting solvent by nitrogen or helium blow-down and then fractionated into separate chemical classes. The preferred separation procedure involves open-column liquid chromatography or HPLC, although liquid–liquid partition has also been used.

Open-column liquid chromatography

A simple glass column is packed with a suitable chromatographic adsorbent and sequential elution with organic solvents separates the PAH or PCDD/PCDF fraction. Careful selection of adsorbents and eluting solvents also allows the fractions to be separated into further subfractions. Separation is effected by the relative affinities of the compounds for the adsorbent and on their relative solubilities in the eluting solvent. Selective use of different solid phases enables removal by adsorption of interfering species, while selective use of different solvents enables removal of adsorbed species from the solid phase. Suitable adsorbents are alumina, chromatographic-grade silica gel, or a lipophilic gel such as Sephadex LH-20. The adsorbent may be modified to produce more selective column packings.

A number of analytical schemes have been developed for the analysis of PAH and PCDD/PCDF. Rosen and Middleton (1955) developed a method for the separation of PAH from environmental samples on activated chromatographic-grade silica gel. Elution with an alkane solvent such as hexane or pentane yields the aliphatic fraction, followed by elution with benzene to yield the aromatic fraction which includes the PAH fraction. The Rosen procedure has been used widely for the separation of PAH from a variety of samples and is the basis for the EPA level 1 analysis of PAH (Jones *et al.*, 1976). There has been some criticism of the alumina, silica gel column chromatography system in that the alumina and silica gel have high adsorptivity which may result in losses of trace constituents.

Alternatively, the use of lipophilic permeation gels such as Sephadex LH-20 has been advocated as an effective separation column adsorbent for PAH. The basis of gel permeation chromatography is the selective retardation of PAH molecules due to the relative degree of penetration into the solvent-filled pores in the gel matrix. Molecules of smaller size permeate the gels much more deeply than larger molecules and are thus retained longer on the column, thereby effecting the separation in reverse order of most chromatographic systems, i.e. elution in order of decreasing molecular size. The method has been used to separate PAH fractions from shale oil and coal oil, diesel exhaust and cigarette smoke. The method also allows for the separation of PAH according to ring size.

Because of the very low concentrations of PCDD/PCDF and the likely presence of other chlorinated organic material, the analytical schemes for PCDD/PCDF which involve liquid chromatography tend to be more involved and complex. For example, the Canadian National Incinerator Testing and Evaluation Programme (NITEP) used a three-column approach of an acid–base column with hexane elution, followed by a silver nitrate column and finally an alumina column which was eluted sequentially with hexane, 1.5% dichloromethane in hexane and finally by 50% dichloromethane in hexane to collect the PCDD and PCDF for subsequent analysis by gas chromatography/mass spectrometry. In another clean-up system for PCDD/PCDF using a dual column system (Ozvacic *et al.*, 1985), the extracted organic mixture was first passed through a column packed with silica, base-modified silica (33% 1 M NaOH), silica, acid-modified silica (44% H_2SO_4) and silica. This column was eluted with hexane, the column retaining chlorinated phenols. The eluant from the column was then concentrated and passed to a second column containing 10% $AgNO_3$/silica and alumina and eluted with pentane. This second column retained PCDDs and PCDFs, while polychlorinated biphenyls (PCBs) and chlorobenzenes were eluted. The PCDD and PCDF which were concentrated in the alumina part of the column could then be eluted from the alumina, which had been separated from the rest of the column, by first washing with 10% CCl_4 in hexane and hexane which were discarded and finally by methylene chloride elution which contained the concentrated PCDD and PCDF fraction.

A number of other schemes have been reported in the literature, all essentially involving liquid column chromatography with various solid phases and the use of selective solvents to remove interfering organic chemical groups and concentrate the PCDD and PCDF. Bushby (1985) and Tiernan (1983) have summarized and discussed several different methods of liquid column chromatography clean-up for the preparation of PCDD- and PCDF-rich fractions.

High Performance Liquid Chromatography

High performance liquid chromatography (HPLC) may be considered as an improvement on the technique of open-column liquid chromatography in terms of speed, reproducibility and efficiency. The HPLC instrument operates with a controlled solvent delivery system which enhances the reproducibility of the flow rate and composition of the solvent mobile phase and allows for automation of the procedure. In addition, for HPLC, all the sample components are separated and eluted sequentially from the column. Open-column chromatography on the other hand separates unwanted components by adsorption to the column stationary phase which is then usually discarded. Such an approach is not valid for HPLC, since the columns are expensive and have to be re-equilibrated before analysis of another sample. The most commonly used HPLC system for organic analysis is reverse-phase chromatography where a non-polar stationary adsorptive phase is used in conjunction with a polar liquid or mobile phase, for example an octadecyl bonded to silica gel stationary adsorptive phase, and water–acetonitrile as the liquid phase. Good separation of PAH and PCDD/PCDF isomers, and particularly alkyl-substituted PAH, is achieved in this mode. In normal-phase HPLC, where the stationary adsorptive phase is polar and the liquid mobile phase is non-polar (for example, silica gel with bonded amine groups as the stationary phase and hexane as the mobile liquid phase), the separations are similar to those obtained with open-column chromatography using silica gel and alumina. However, much higher efficiencies and speed are achieved.

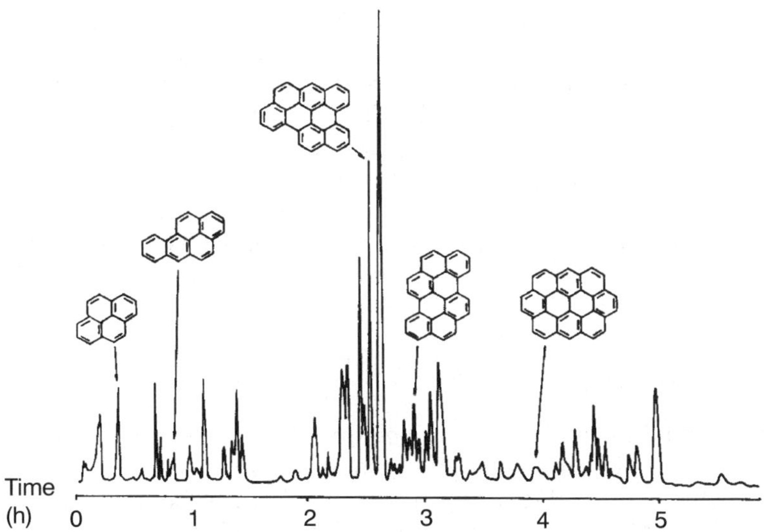

Fig. 10.4 High-pressure liquid chromatography of a high-molecular weight PAH fraction extracted from carbon black..

The resolution of compounds attainable by HPLC is about an order of magnitude lower than may be achieved with capillary column gas chromatography. Consequently, HPLC is most often used as a high-resolution pre-fractionation technique for complex mixtures before analysis by capillary column gas chromatography. However, HPLC is superior to gas chromatography for the analysis of high-molecular weight compounds such as PAH with high boiling points which may not be eluted from a capillary column at the temperatures currently achievable with gas chromatographs, while still maintaining stability of the stationary adsorptive phase. Figure 10.4 shows the analysis of high-molecular weight PAH in carbon black by HPLC (Peaden *et al.*, 1980). Ultra-violet absorption (UV) and fluorescence detectors used in HPLC are extremely sensitive and selective for the determination of PAH. In addition, HPLC has the advantage of being non-destructive, allowing fractionation of the organic micropollutants for further analysis. There are a variety of stationary adsorptive phases available for the HPLC column and these, coupled with suitable solvents, provide unique selectivity for PAH and heterocyclic PAH. These have been reviewed extensively (Lee *et al.*, 1981). HPLC has been used both as a final analytical step and a pre-fractionation system before gas chromatographic analysis in a number of applications, for example carbon black, air particulates, automobile exhaust and fossil fuels. HPLC with selective columns has been used to separate nitrogen-containing PAH, nitro PAH and to separate PAH by ring size.

HPLC analysis has also been used for the fractionation of PCDD and PCDF from the organic sample matrix (Tong *et al.*, 1984). A benzene extract was fractionated using a silica HPLC column and gradient elution consisting of a programmed sequence of controlled solvent gradient elution of, *n*-hexane, methylene chloride, acetonitrile, methylene chloride and *n*-hexane to isolate a PCDD/PCDF-rich fraction. The combination of open-column liquid chromatography and HPLC has been used to isolate PCDD/PCDF fractions (Lamparski and Nestrick, 1980). Preliminary fractionation to produce a PCDD/PCDF-rich fraction is carried out using a three-open-column system followed by reverse-phase HPLC fractionation to produce sub-fractions of four-, five-, six-, seven- and eight-chlorine-substituted PCDD/PCDF.

Liquid–liquid partition

Liquid–liquid partition has also been used to isolate a PAH-rich fraction from the sample extract using a second immiscible solvent. If the partition coefficients for PAH differ from those of the other materials present, they will be preferentially concentrated in one of the layers, and multiple partitioning is usually required. A number of partition solvents and combinations of solvents have been applied to isolate the PAH fraction (Lee *et al.*, 1981). Recommended solvents with high partition coefficients for PAH are, nitromethane, dimethylsulphoxide (DMSO) and dimethylformamide (DMF). Some analyses of PCDD/PCDF samples have involved a liquid–liquid partition step as part of the sample clean-up, which also involves in addition either HPLC or open-column chromatography (Choudhary *et al.*, 1983).

10.5.3 Final analysis

Fractions collected in the preliminary separation stage by liquid column chromatography, HPLC or liquid–liquid partition are subsequently concentrated by removal of solvent by nitrogen or helium blow-down. They are then analysed most commonly by capillary column gas chromatography, usually coupled to a mass spectrometer. The capillary columns are made of glass or fused silica with a thin coating of stationary adsorptive phase on the inside surface. Columns are between 10 and 25 metres in length. Suitable mobile gas phases are nitrogen, helium and hydrogen. A variety of stationary phases have been used in capillary column separations of PAH but, because of their selectivity and high temperature stability, SE 52 (methylsilicone gum with 5% phenyl groups) and SE 54 (methylsilicone gum with 1% vinyl and 5% phenyl groups) are mostly used. The resolution of PAH using capillary column gas chromatography is illustrated in Fig. 10.5, which represents the PAH fraction of coal combustion products using an SE 52 column and hydrogen carrier gas (Lee *et al.*, 1977). Columns used for PCDD/PCDF analysis range from the highly polar Silar 10C (a polycyanopropylsiloxane silicone phase) to the non-polar DB-5 (a diphenyldimethyl polysiloxane bonded phase). Other types of column have included, OV17, SP2200, SP2330, SE54, etc., the names referring to the commercial designations. Bushby (1985) and Tiernan (1983) have reviewed the different types of capillary columns available for PCDD/PCDF analysis. Essentially, the majority of columns will achieve suitable resolution of the PCDD and PCDF isomers with variations occurring due to the efficiency of separation of the isomers within a group or from other chlorinated aromatic hydrocarbons.

The identification of organic micropollutants from gas chromatographic traces is by way of internal standards and retention indices. Standards are assigned a retention index and calculation of relative retention indices between standards and the use of tables enables the identification of PAH, PCDD and PCDF. The identifications are usually confirmed using capillary column gas chromatography/mass spectrometric detection. The effluent from the gas chromatograph is passed directly to the mass spectrometer for mass spectral analysis, in most cases using electron impact mass spectrometry. The use of mass spectral libraries with computer search facilities enable rapid identification of organic species. Electron

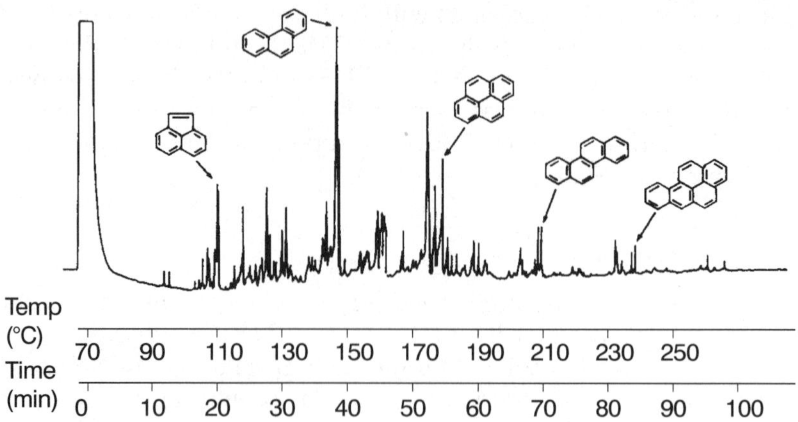

Fig. 10.5 Capillary column gas chromatogram of the PAH fraction of coal combustion products.

impact mass spectra of PCDD and PCDF exhibit intense molecular fragmentation ions and in most instances monitoring the molecular ions of the PCDD and PCDF will distinguish these compounds from other chlorinated hydrocarbons. However, some polychlorinated biphenyls and chlorinated diphenyl ethers can fragment to yield ions which may interfere with the determination of PCDD and PCDF. The use of mass spectrometry electron impact alone is usually not sufficient for positive identification and retention indices and operator experience are also required. Selective detectors are available for gas chromatography analysis which enable specific detection of heterocyclic PAH. For example, PANH may be detected with the use of a nitrogen phosphorus detector (NPD) which is specific for nitrogen compounds, a flame photometric detector (FPD or SSD) is used for sulphur-specific detection, and electron capture detection (ECD) is specific and sensitive for certain PAH and halogenated compounds. Figure 10.6 shows the use of a three-way parallel effluent splitter to determine the total PAH (FID), PANH (NPD) and PASH (SSD) in diesel fuel (Williams *et al.*, 1986).

Where further distinction between isomers is required, interfering species are suspected, or concentrations are very low, gas chromatography/mass spectrometric detection with selective ion monitoring (SIM) is used. The mass spectrometer is tuned to record the response of a particular characteristic fragmentation ion or set of ions rather than scan through the full mass spectral range. Figure 10.7 shows the gas chromatographic separation with mass spectrometric detection using selective ion monitoring for the tetra-CDD, penta-CDD, hexa-CDD, hepta-CDD and octa-CDD using single ion detection with m/z 320, 354, 388, 424 and 458 respectively (Buser and Rappe, 1984).

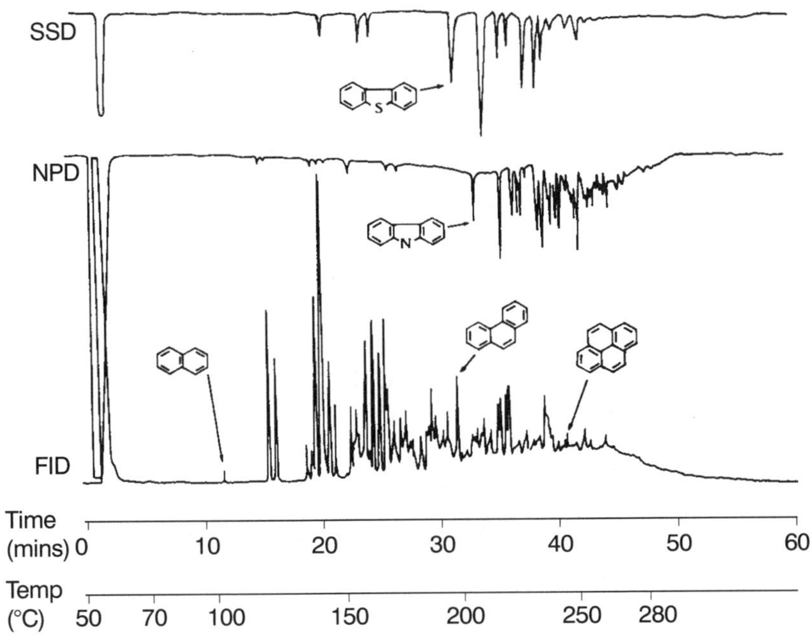

Fig. 10.6 Gas chromatogram of diesel fuel PAC fraction with flame ionization detection (FID), nitrogen- (NPD) and sulphur- (SSD) selective detection.

10.6 SUMMARY

In summary, the most widely used scheme of sampling of the organic micropollutants PAH, PCDD and PCDF from combustion streams are the use of glass-fibre filters to trap the solid particulate with adsorbed organic material and a back-up polymeric resin trap for the gaseous organic compounds. The most common sampling system is the EPA Method 23. Water condensates which contain the organic micropollutants may also be required to be sampled. Analysis involves extraction of the organic compounds from both the glass-fibre filters and polymeric resin by Soxhlet extraction using a suitable organic solvent such as methylene chloride or benzene : methanol. Extraction of the water condensate is by liquid–liquid partition with an immiscible organic solvent. The extracts from the filter, resin and water condensate may be analysed separately or together for a bulk analysis. The extracts which contain many classes of organic compounds in addition to PAH, PCDD and PCDF require preliminary separation to isolate the organic micropollutants of interest. This is most readily achieved either using open-column liquid chromatography with alumina and/or chromatographic silica gel as the chromatographic adsorbent which may be modified, or using HPLC. HPLC provides a large

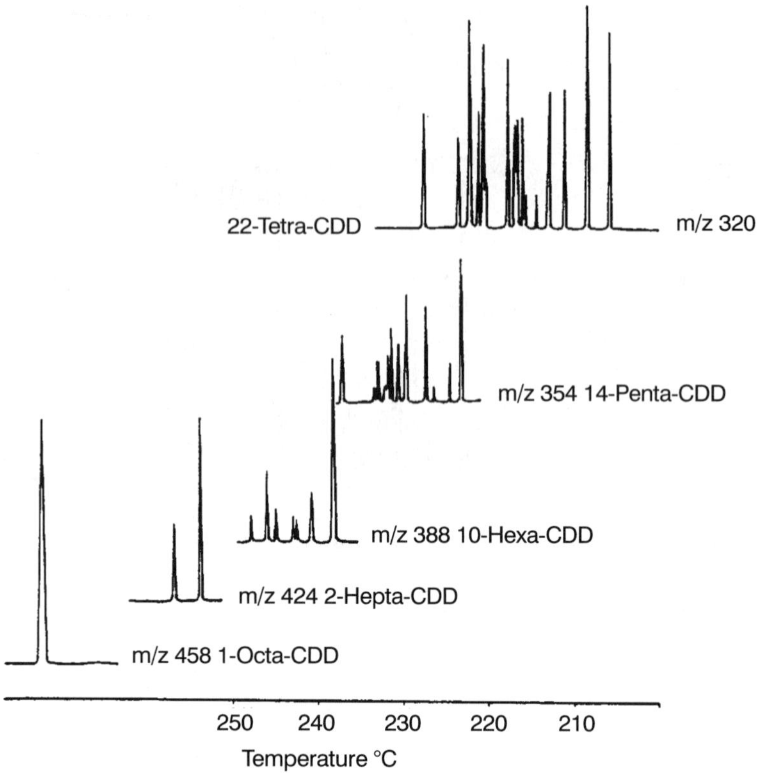

Fig. 10.7 Gas chromatogram/mass spectrogram of a composite sample of PCDD.

variety of solid adsorptive stationary phases leading to separation of the organic micropollutants from the extract and also fractionation of the PAH into subclasses such as PANH, PASH, etc. HPLC is also a rapid and efficient method of separation. The separated fraction is then analysed by capillary column gas chromatography coupled to a mass spectrometer for the mass spectral analysis of the organic compounds coupled with retention index data.

A recent (1996) PCDD/PCDF sampling and analysis standard has been proposed (CEN EN1948 – Stationary Source Emissions – determination of mass concentration of PCDDs/PCDFs – Parts 1 to 3). The standard essentially follows the strategy outlined in this chapter, dealing with sampling, extraction and clean-up and finally identification and qualification.

REFERENCES

Buser, H.R. and Rappe, C. (1984) *Analytical Chemistry*, **56**(3), 442–8.

Bushby, B.R. (1985) *A review of the analytical procedures used for polychlorinated dibenzo-p-dioxins and polychlorinated dibenzofuran determinations in combustion effluent*. Warren Spring Laboratory, Report No. LR 613 (CS)M. HMSO, London.

Choudhary, G., Keith, L.M. and Rappe, C. (eds) (1983) *Chlorinated Dioxins and Dibenzofurans in the Total Environment*, Butterworth, London.

Hagenmaier, H., Brunner, H., Haag, R., Kraft, M. and Lutzke, K. (1988) in *Incineration of Municipal Waste* (ed. R.B. Dean), Academic Press, London.

Jones, P.W., Graffeo, A.P., Detrick, R., Clarke, R.A. and Jacobsen, R.J. (1976) *Technical Manual for Analysis of Organic Materials on Process Streams*, EPA-600/2-76-072.

Lamparski, L.L. and Nestrick, T.J. (1980) *Analytical Chemistry*, **52**, 2045–54.

Lee, M.L., Novotny, M. and Bartle, K.D. (1981) *Analytical Chemistry of Polycyclic Aromatic Compounds*, Academic Press, New York.

Lee, M.L., Prado, G.P., Howard, J.B. and Hites, R.A. (1977) *Biomedical Mass Spectrometry*, **4**, 182.

Lustenhouwer, J.W.A., Olie, K. and Hutzinger, O. (1980) *Chemosphere*, **9**, 501.

Ozvacic, V., Wang, G., Tosine, H., Clement, R.E. and Osborne, J. (1985) *Journal of the Air Pollution Control Association*, **35**(8), 849–55.

Peaden, P.A., Lee, M.L., Hirata, Y. and Novotny, M. (1980) *Analytical Chemistry*, **52**, 2268.

Rosen, A.A. and Middleton, F.M. (1955) *Analytical Chemistry*, **27**, 790.

Schuetzle, D. and Perez, J.M. (1981) 74th Annual Meeting Air Pollution Control Association, Philadelphia.

Tiernan, T.O. (1983) in *Chlorinated Dioxins and Dibenzofurans in the Total Environment* (eds G. Choudhary, L.M. Keith and C. Rappe), Butterworth, London.

Tong, H.Y., Shore, D.L., Karasek, F.W., Helland, P. and Jellum, E. (1984) *Journal of Chromatography*, **285**, 1830–42.

Williams, P.T., Bartle, K.D. and Andrews, G.E. (1986) *Fuel*, **65**, 1150.

11

Electrochemical and other non-optical techniques

Andrew G. Clarke and Ian Watson

11.1 INTRODUCTION

Electrochemical devices have found very wide application in the field of toxic gas monitoring in the context of industrial hygiene. Until about 10 years ago they had not been as widely used for stack gas or exhaust monitoring with the exception of the zirconia sensor for oxygen. However, there are now many portable flue gas analysers based on these cells offering the potential to measure up to about six gases in one unit. There are no problems with sensitivity to the pollutant gases found in flue gases since the concentrations are generally higher than in ambient factory atmospheres. However, there may be more severe problems of selectivity, resistance to contamination and operational life than would be encountered in industrial hygiene applications.

The great attraction of electrochemical sensors is the potential for low costs. The sensors themselves are quite cheap and the overall cost of analysers can be well below those based on optical or other more sophisticated technologies. This comment applies equally to the non-specific combustible gas detector and semi-conductor sensors.

This chapter starts by examining the methods for analysis of oxygen, including the paramagnetic analyser, as well as electrochemical devices. Other low- and high-temperature cells are then considered. For convenience, approaches based on mass spectrometry are also grouped here since they do not fall conveniently into other chapters.

11.2 METHODS FOR OXYGEN

The measurement of oxygen in gas mixtures has been extensively reviewed by Kocache (1986).

11.2.1 Solid-state zirconia sensors

For details of this technique, the reader is referred to reviews by Kocache (1986), Moseley and Tofield (1987), and Moseley *et al.* (1991a).

The ceramic material zirconia (ZrO_2), when heated above about 600°C, becomes an oxygen ion conductor. Most sensors make use of yttria-stabilized zirconia in which the presence of the additional yttrium ion contributes oxygen ion vacancies in the crystal lattice. It then becomes possible to set up an oxygen gas concentration cell in which a difference in oxygen partial pressure across the element generates an e.m.f. across the electrodes. The oxygen partial pressure on the reference side of the cell is known and therefore the oxygen partial pressure of the flue gas side can be calculated. Figure 11.1 shows a zirconia element equipped with platinum electrodes, thermostatic heating and a source of air as a reference gas. The e.m.f., E, of the cell is given by the usual Nernst equation:

$$E = \frac{RT}{4F} \ln \left[\frac{p_1}{p_2}\right]$$

where R is the gas constant, T the absolute temperature, F is Faraday's constant, p_1 the partial pressure of oxygen in air, and p_2 is the partial pressure of oxygen in the flue gas.

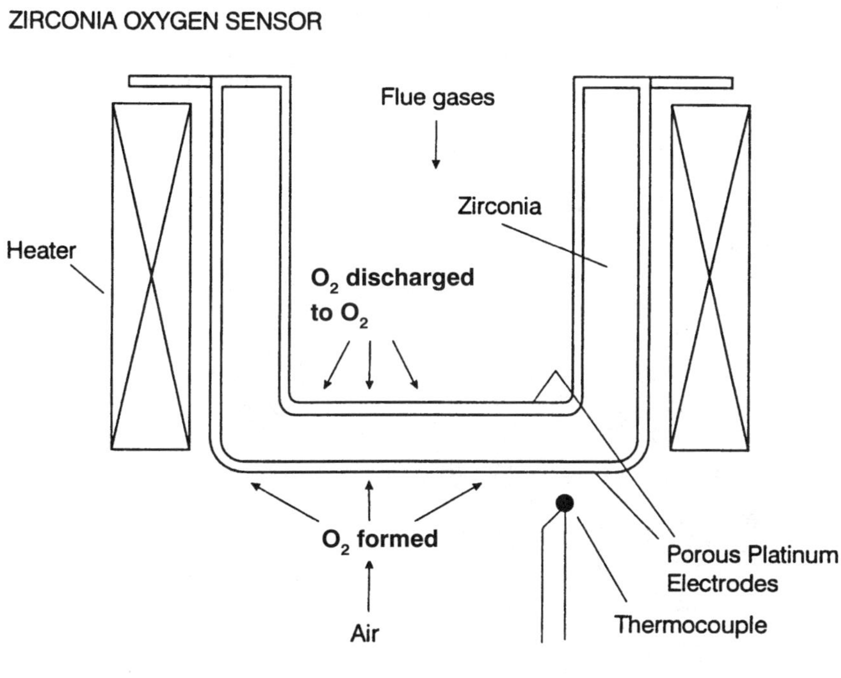

Fig. 11.1 Schematic diagram of the principle of operation of the zirconia oxygen sensor.

Zirconia cells are generally operated at a temperature in the region of 750°C and consequently can be inserted into flue gases approaching this temperature. Various configurations are possible. The zirconia may be in test-tube shape, as a disk in the centre of a zirconia tube, or as a flow-through tube. Sensors used for automobile exhaust gas fuel–air ratio control are smaller but operate on the same principle. The control of the temperature is critical to the accuracy and the sensor must be surrounded by a heater in such a way that temperature gradients in the cell are minimized.

A common form of this type of oxygen analyser comprises a probe-mounted zirconia element inserted into the flue duct (Fig. 11.2). A porous sinter filter may be used to protect the element. Electrical signals from the probe are taken to an electronics unit where O_2 concentration is calculated, calibration adjustment is controlled and microprocessor-based self-diagnostics may be incorporated. An alternative arrangement is to mount the sensor and its heating system just outside the flue. Rack-mounted units which can be housed alongside other gas analysers are also available and these can be used for automobile exhaust gas analysis.

A recommended safety precaution against any form of unwanted ignition of unburned fuel – particularly in gas-fired applications – is the use of a flame trap. This consists of a sintered steel cylinder attached over the zirconia sensor and acts on the same principle as the Davy safety lamp. Some systems require permanent reference air to be pumped to the probe, but other analysers include a method of using convection currents

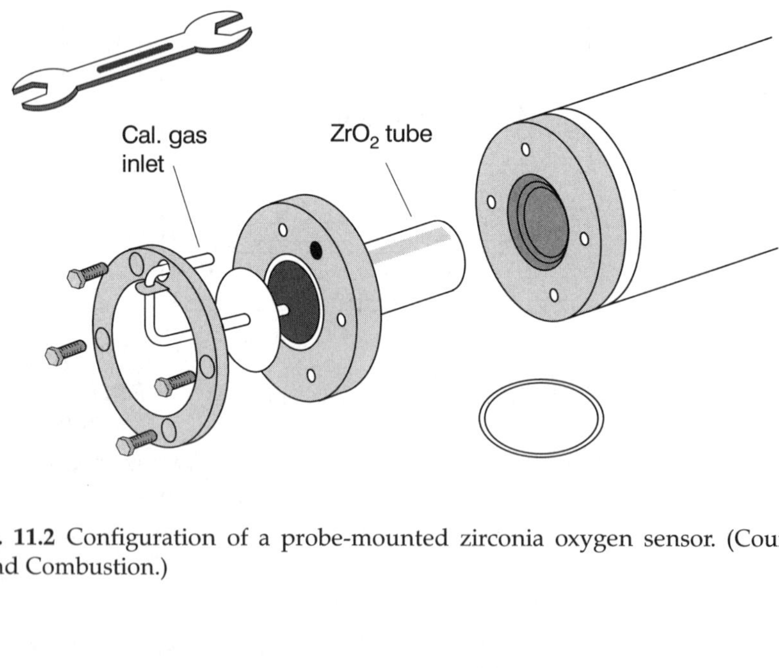

Fig. 11.2 Configuration of a probe-mounted zirconia oxygen sensor. (Courtesy Land Combustion.)

to draw in normal fresh air eliminating the need for a separate reference source.

Zirconia sensors have fast response and low maintenance requirements. Sensor replacement is simple. The reading is however the *wet-gas oxygen level.* Correction to standard dry gas conditions requires the dry-gas level. Correction for water vapour is therefore necessary. Some manufacturers are actually using dual wet and dry gas zirconia systems to calculate the water vapour content.

11.2.2 Low-temperature electrochemical cells

The basic electrochemical cell which has a cathode, an anode and an aqueous electrolyte is well known. When an e.m.f. is applied to it, oxygen evolves at the anode and hydrogen at the cathode. Its operation in the reverse mode is the basis of galvanic oxygen sensors (Kocache, 1986; Moseley *et al.*, 1991b). When an oxygen concentration is established round the cathode which is usually a precious metal (e.g. Pt) and an appropriate electrolyte such as KOH is used, the anode, which is usually a base metal (e.g. Pb), is oxidized to a hydroxide with the generation of an electronic charge:

$$O_2 + H_2O + 4e^- \rightarrow 4OH^-$$
$$Pb \rightarrow Pb^{2+} + 2e^-$$
$$Pb^{2+} + 2OH^- \rightarrow Pb(OH)_2$$

If a load is connected to the cell, the current flowing would be proportional to the partial pressure of oxygen in the mixture. In practical cells the gas sample is led through a thin solid membrane (e.g. Teflon) that allows the preferential diffusion of oxygen molecules to a large-area cathode onto which the oxygen is absorbed and moves into contact with the electrolyte where it is ionized. Figure 11.3(a,b) illustrates one design of cell in which the access of oxygen is limited by a capillary restriction. The oxygen concentration at the cathode is very low and the current is therefore limited by the diffusion rate.

For most purposes the oxygen cell may simply be calibrated in ambient air. It is usual to carry out all other span and range checks with test gas mixtures using a carrier gas of nitrogen. If other carrier gases are used, care should be taken to compensate for the different molecular weight of the carrier gas.

The specifications for instruments using these cells are designed to accept continuous gas temperature to the cell in the range –15 to +40°C for a pressure of 0.9–1.1 atmospheres over full humidity range.

Such oxygen detectors are commonly used in conjunction with other low-temperature sensors described below.

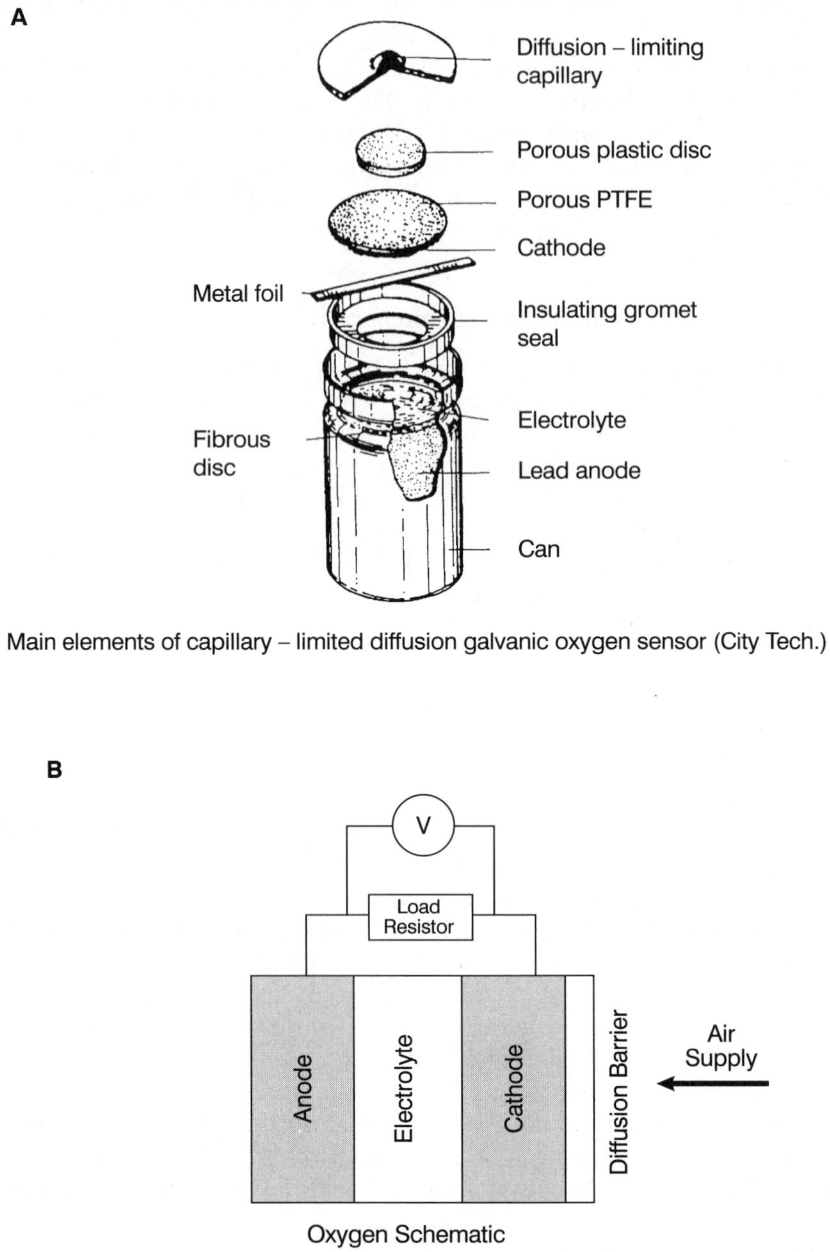

Fig. 11.3 (a) Main elements of a capillary-limited diffusion galvanic oxygen sensor. (b) Schematic electrical components of an oxygen cell. (Courtesy City Technology.)

11.2.3 Paramagnetic analyser

Although not strictly an electrochemical method the determination of oxygen by making use of its paramagnetism is included here for comparison with other methods (Kocache, 1986; Moseley *et al.*, 1991c).

Oxygen is unusual in having unpaired electrons and this gives it its property of paramagnetic susceptibility. NO is one of the few other common gases with this property but because of its low concentration in typical flue gases it is not normally an interferent. Paramagnetic substances are attracted to the strongest part of a magnetic field. A typical analyser system based on this principle is illustrated in Fig. 11.4(a,b). A test body, shaped as a dumb-bell, and consisting of two glass microspheres filled with nitrogen, is suspended in a magnetic field. A small mirror is attached to the centre of the test body. A feedback coil round the test body provides a means of moving it when an electric current is allowed to pass through it. The system is housed in a gas-tight chamber with an optical window and a means of admitting gases into the chamber. The movement of the test body is detected by an optical system (Fig. 11.4(b)) and the signal from the photocells is used to initiate a current in the feedback coil restoring the dumb-bell to its equilibrium position. The current necessary to maintain this position depends on the oxygen content of the gas.

Paramagnetic oxygen analysers are reliable devices providing they are supplied with clean, dry gas. They are commonly used as stand-alone portable analysers, rack-mounted analysers and are sufficiently compact to be incorporated in a single unit with infra-red analysers for other gases.

11.3 OTHER LOW-TEMPERATURE ELECTROCHEMICAL CELLS

A very wide range of cells have been developed (Moseley *et al.*, 1991b), originally for use as toxic gas sensors. Common pollutants that can be measured include CO, SO_2, H_2S, NO, NO_2, Cl_2 and HCl. The cells contain three electrodes – a sensing electrode, a counter electrode and a reference electrode separated by a thin layer of electrolyte (Fig. 11.5). Gas diffusing to the surface of the electrode is either oxidized (e.g. CO, H_2S, SO_2, HCN) or reduced (NO_2 or Cl_2). The selectivity to different gases is determined by the specific electrode material. The counter electrode built into the cell acts to balance out the reaction at the sensing electrode by reducing O_2 in air to water. A gaseous diffusion barrier ensures that the flow of gas to the sensing electrode does not exceed its electrochemical activity and also results in a direct response to volume concentrations rather than to a partial pressure.

Toxic gas cells will usually operate to specification over the range –5 to +40°C. It will be obvious that these cells cannot be inserted directly into a

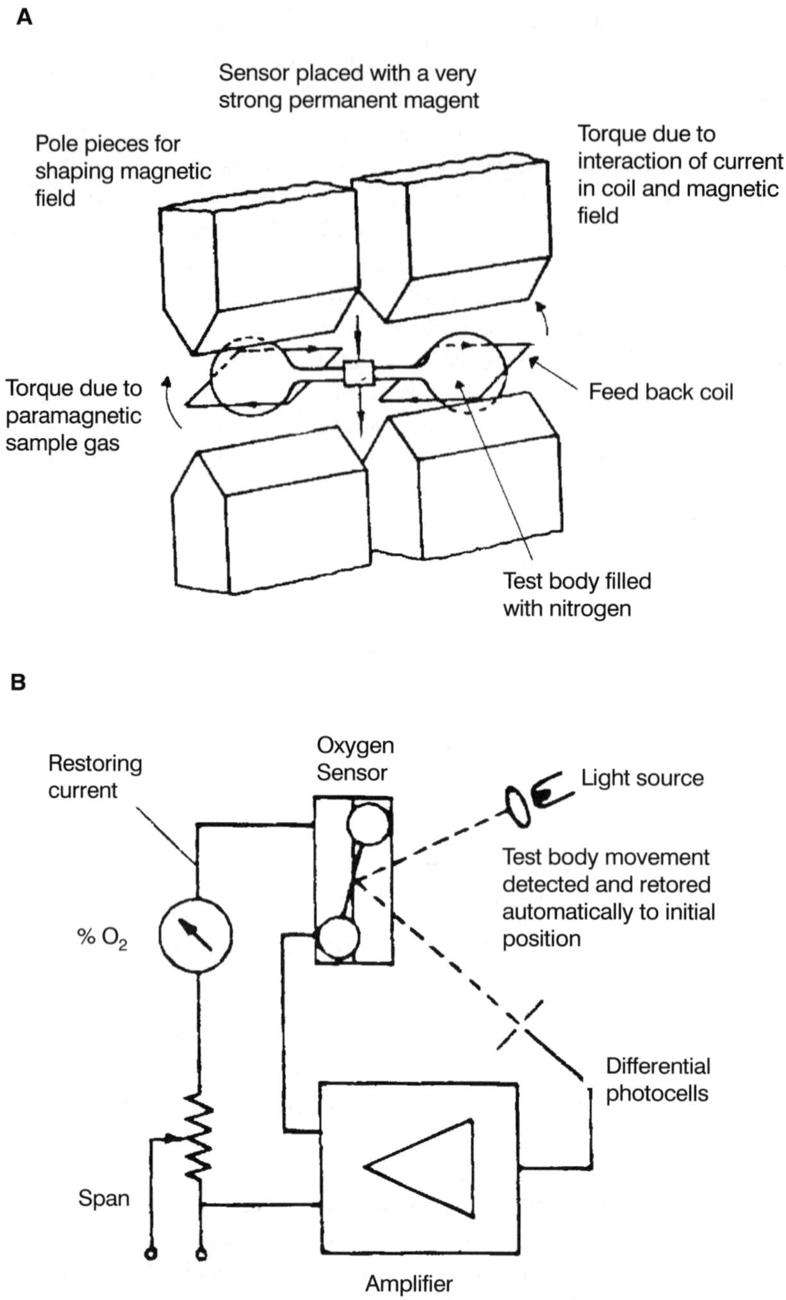

Fig. 11.4 (a) Arrangement of test body and magnets in a paramagnetic oxygen analyser. (b) The magnetic auto-balance system with current feedback. (Courtesy Servomex.)

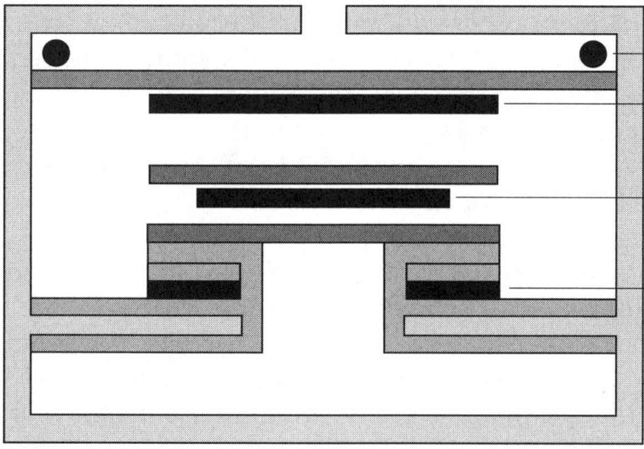

Fig. 11.5 Structure of a toxic gas CiTiceL cell. (Courtesy City Technology.)

flue gas stream and it is necessary to introduce a sampling system which cools the flue gases before entry into the measurement system. Hence the major use for these cells as flue gas monitors has been to incorporate them into portable instruments. There are many significant points that instruments incorporating these cells should have included in their design such as:

- The gas sample system should be designed with appropriate filters to remove highly corrosive constituents and particulate matter.
- Sample gas lines should be in materials which do not absorb the gas components. This is particularly important for gases such as SO_2 and NO_2 where PTFE is normally recommended.
- Condensing conditions at the cell must be avoided.
- If used intermittently, cell life is improved by purging toxic gas cells with clean air after use.

Table 11.1 Cross-sensitivity data for electrochemical cells used in flue gas analysis (City Technology.)

Cell	Cell Reference	CO	H_2S	SO_2	NO	NO_2	H_2	Cl_2	HCN	HCl	C_2H_4
CO low	3F/F	100	<1	<1	<1	<1	<60	N/D	<5	<1	<1
CO high	3M	100	ND	ND	ND	ND	ND	ND	ND	ND	ND
SO_2	3SF	<1	≈200	100	1	<–125	<3	≈–80	≈30	ND	≈180
NO	3NF/F	0	0	0	100	1	0	0	0	<5	0
NO_2	3NDH	0	≈–20	<1	0	100	<1	≈90	ND	<1	0

ND, not detected

The possibility of cross-sensitivity of a particular gas being monitored to other flue gas constituents needs to be considered. Many of the cells have very small cross-sensitivities to other gases and on flue gas monitors these cross-sensitivities are negligible. However, some cells show considerable cross-sensitivity to other possible flue gas constituents (e.g. SO_2 and H_2S; see Table 11.1). In such circumstances the supplier of the cell or the company incorporating the cell into their equipment will make provision to filter out offending gas before introduction to the affected measuring cell.

A typical portable unit for flue gas monitoring is illustrated in Fig. 11.6. The clean dry flue gas enters the unit and is distributed to the individual sensors from a manifold. However, to avoid the influence of SO_2 and NO_2 on the CO cells the gas is then passed through a chemical filter before accessing the CO cells. In this particular unit, two CO cells are used to cover the very wide range of possible concentrations (from 100s of ppm to % levels). The appropriate range cell is selected to avoid, for example, very high concentrations saturating the low range cell. Carbon dioxide is not normally measured directly in these systems but a value may be computed as part of a built-in software package which enables combustion efficiency to be evaluated. The monitoring of temperature, pressure and gas flow may also be offered.

There are a large number of manufacturers now offering such systems and they are very attractively priced. However, they must be used with particular attention to the bulleted points listed above. Frequently, they are sold with minimal sample conditioning arrangements – an unheated sample line, a filter and a moisture droplet trap. Getting condensation at a cell can be highly damaging and the possibility of further cooling and droplet formation after entry into the instrument needs to be considered. Sample losses of O_2, CO and NO should not be a problem, but minor losses of NO_2 and SO_2 may occur. Sulphuric and hydrochloric acids can be very detrimental if they penetrate the filtration system. Some manufacturers may not include any chemical filtering system and this can cause cross-sensitivity problems with some gases as described above. Suppliers may not provide information about cross-sensitivity unless such information is specifically requested. If operated at high concentrations for long periods, the cells cease to respond to the component to be measured. They are therefore best operated for short periods (e.g. <1 hour), followed by a clean air purge (e.g. 15 minutes). They are thus not suitable for long-term, unattended operation unless specifically tailored to that application. Figure 11.7 illustrates a system designed for $CO/NO/O_2$ detection in gas-fired installations or coal mills which switches automatically between pairs of sensors mounted in parallel.

Other points which are worth considering by any user of such equipment are:

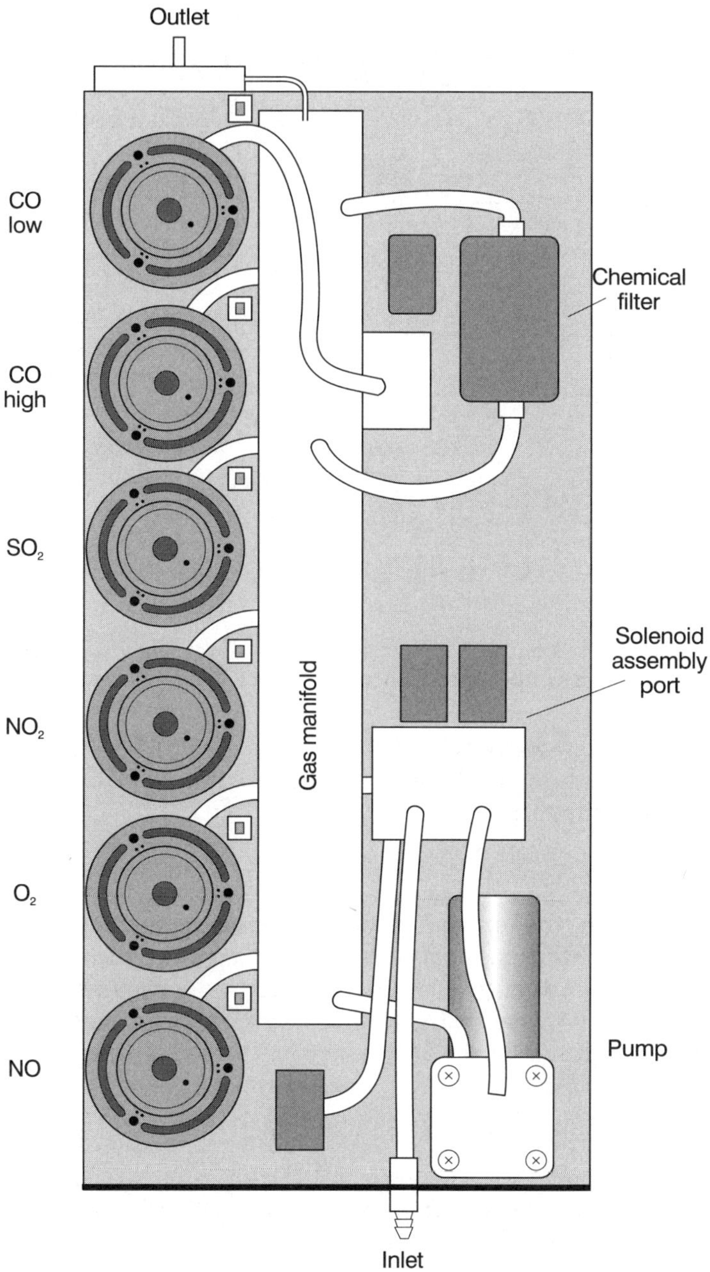

Fig. 11.6 Layout of sensors in a portable flue gas analyser. (Courtesy Land Combustion.)

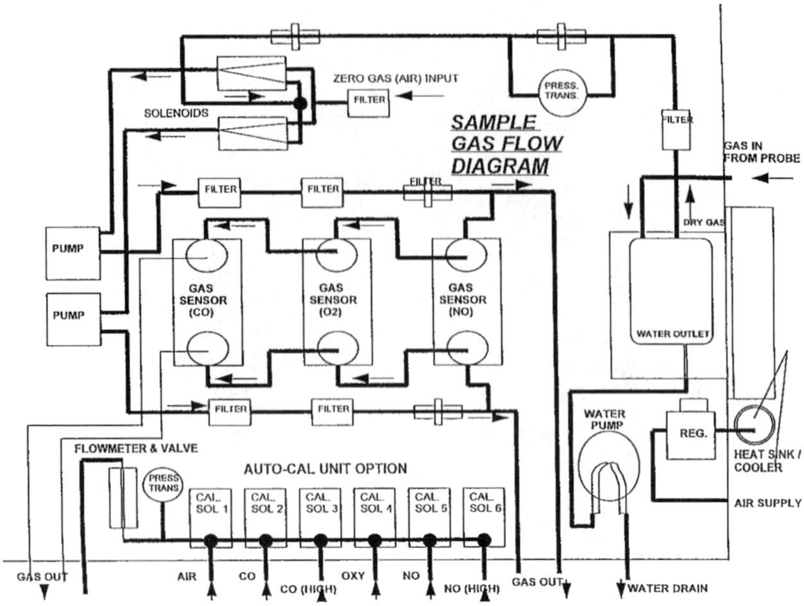

Fig. 11.7 Monitor for gas-fired installations or coal mills with pairs of sensors switched between sample and clean air. (Courtesy Land Combustion.)

- How easily can the cells be changed?
- Can confirmation of calibration of the instrument – and any re-calibration – be easily carried out?
- Will the software and data presentation meet the needs of legislation now or in the future?

In conclusion, the ambient temperature electrochemical cell is widely used in portable flue gas analysers and in selected applications for continuous gas emissions. It gives a stable and reproducible output over a long time period if used correctly, and can be checked easily for calibration accuracy against standard gases.

11.4 HIGH-TEMPERATURE ELECTROCHEMICAL CELLS

The *in situ* zirconia oxygen sensor previously discussed allows a user to measure concentrations of flue gas in a duct directly. However, ambient temperature electrochemical sensors can only be used for flue gas measurements in conjunction with a sampling system. An obvious development was to produce electrochemical sensors which could be mounted in a probe giving all the advantages of the *in situ* principle of measurement. There are not many such systems and the ones described here were developed by Land Combustion to make measurements of major flue gas constituents such as SO_2 and NO_x. It is possible continuously to monitor these gases with electrochemical cells inserted directly into flue gas streams up to 260°C.

The construction principle of this type of cell is shown in Fig. 11.8. It consists of two major components:

1. The electrolyte used in the assembly is phosphoric acid which at room temperature is in a semi-liquid form but which when heated in use becomes less viscous and quite corrosive.

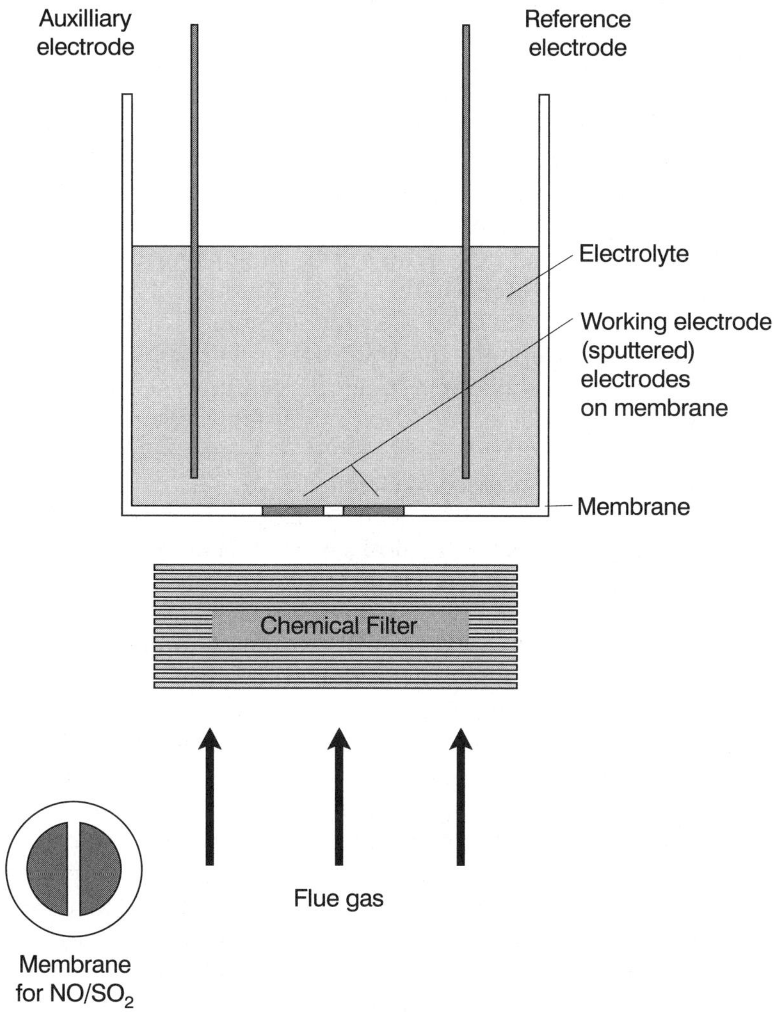

Fig. 11.8 High-temperature electrochemical cell. (Courtesy Land Combustion.)

2. A polymer film spluttered with a metallized surface to form a semi-permeable membrane. The chosen metals determine the gases to which the sensor is sensitive.

A selected milli-voltage is applied across the auxiliary electrode and the working electrode. This voltage has been determined and set at an optimum level where the current flow in the circuit responds to the rate of the gas permeating through the membrane. The current flow then gives a change proportional to the amount of gas present. The reference electrode is introduced to ensure the electrolyte is kept at ground potential, and excess or deficiency of electrons in the acid is automatically corrected by the control of this component.

Life and response of the sensor is dually dependent on the thickness of the polymer and the thickness of the metallized surface. A thicker polymer generally increases the life of the sensor, and the thickness of the metallized surface determines the response time (the thicker the metal, the slower the response). By careful control the membrane has been designed to maximize the response and life of the sensor so as to satisfy both the requirements of the industry and that of the emission regulating bodies.

The process used to deposit the correct amount of metal accurately onto the polymer is a form of spluttering machine. This deposits the chosen metals by bombarding the polymer surface with a radiation beam, the resulting action being similar to a static charged process. Control over the amount of deposit is determined by a crystal oscillator which is also exposed to the same radiation and results in a change to the crystal frequency which can then be used as a control.

Three instruments have so far been developed from the application of the high-temperature electrochemical sensor which are used for NO and NO/SO_2 monitoring. These have been applied in the UK, in Europe and in the USA over a wide range of industrial applications.

One particular set back in the early development of the sensors was the adverse effect that HCl had on the SO_2 function of the sensor, which resulted in a very short life expectancy. This has been overcome by the addition of a special chemical filter which absorbs any HCl before it reaches the sensor and has any harmful effect. Obviously, chemical filters become saturated and have to be replaced periodically to avoid damage to the sensor. Moreover, the manufacturer does not recommend the use of the NO/SO_2 sensor in applications with a high HCl concentrations, such as incineration plants and some coal-fired burner installations.

Experience with the NO sensor has been good and no other components of flue gases have been found to affect the operation. The only limiting factor so far has been the flue gas temperature. Due to the positive results from the NO sensor, a further product has been derived retaining the response, accuracy and stability of the NO instrument at a reduced cost that will allow NO monitoring to be applied to a larger proportion of the industry.

Such a cell installed on a probe with back-up electronics allows operators direct measurements of flue gas concentrations and systems can also be designed which allow for automatic confirmation of the calibration accuracy of the cell by reference to calibration gases – all without the need to remove the system from the plant or any operator involvement other than to check the results at some convenient time.

11.5 SEMI-CONDUCTOR SENSORS

Solid-state semi-conductor sensors (Moseley *et al.*, 1991d) are offered for toxic gas monitoring and flammable gas alarms by some companies. Their applications are complementary to the wet cells previously described. To date, there has not been a significant amount of use for flue gas monitoring.

Tin IV oxide (SnO_2) is non-stoichiometric in that it is deficient in oxygen atoms. Charge neutrality is maintained by the presence of some tin II ions (Sn^{2+}) in place of Sn^{4+} and these act as electron donors, so that the material becomes an n-type semiconductor.

If a bead of tin oxide is heated in clean air, oxygen is absorbed onto the surface layers, accepting electrons and becoming O_2^-, O^- or O^{2-}, until an equilibrium condition is achieved for the particular temperature involved. If a pair of contacts is applied to the bead, a characteristic conductance can be measured which depends on temperature and O_2 concentration. If a combustible contaminant is present however, it will also absorb, then react with some of the oxygen, so releasing electrons from the oxygen which become available for conduction. This means that the bead conductance rises until equilibrium is again attained. Consequently, the conductance can be used as a measure of the contaminant concentration provided that the oxygen and water vapour concentrations are constant. In the context of ambient air monitoring, the O_2 of the atmosphere is, of course, constant and the water vapour concentration is low but this is not the case in flue gases. For an oxidizing gas the reverse effect occurs and the electrical resistance rises.

The earliest sensors were of the bead type with heating of the sensor achieved by an internal heating coil (Fig. 11.9(a,b)). More recent planar designs make use of thick film printing technology achieving significant reductions in size, power requirement and reproducibility (Fig. 11.9(c)). A semi-conductor gas sensor can be made partially selective to hydrocarbons (such as CH_4) by operating it at a high temperature, or to CO by operating at lower temperature. Sensors may be 'doped' to increase selectivity. Using $PdCl_2$ increases sensitivity to methane, while Pt enhances CO response.

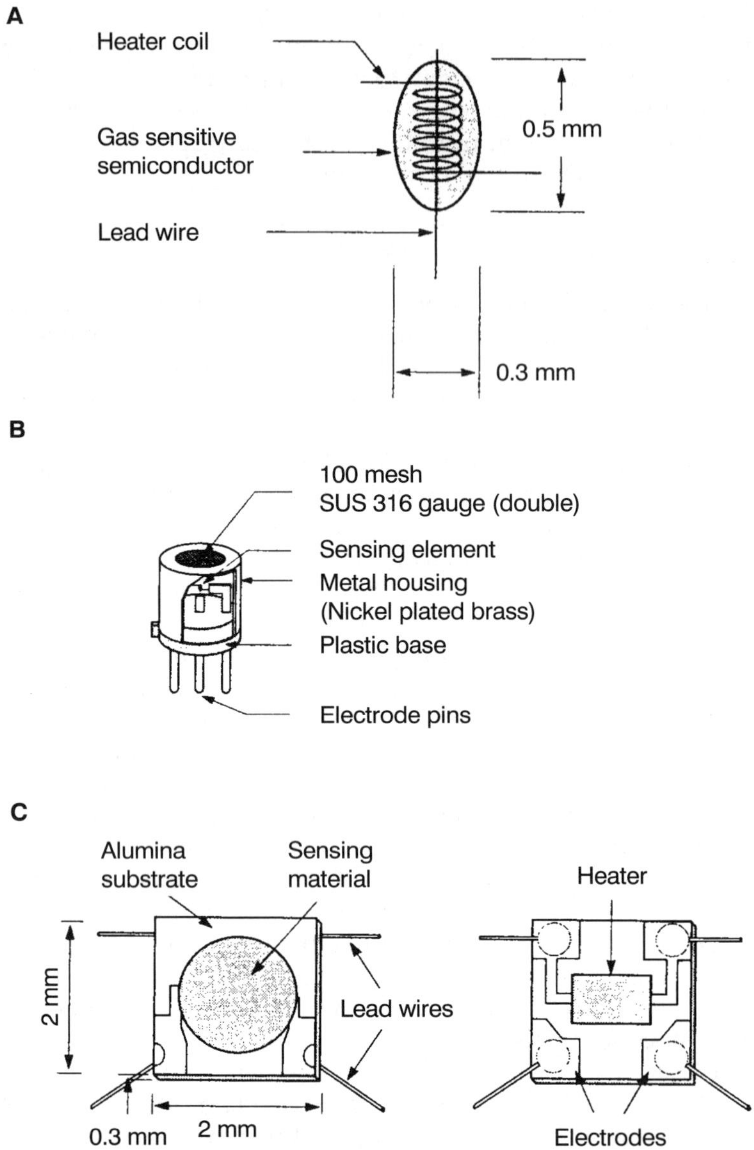

Fig. 11.9 Semi-conductor gas sensors. (a) Bead-type sensing element. (b) Sensor mounting. (c) Planar-type sensing element. (FiS Inc., courtesy Envin Scientific Products.)

For stack or exhaust gases the primary application would be for CO and other combustible gases in competition with the type of sensors described in the next section. For ambient work, semiconductor sensors are also available for NH_3, H_2S, etc.

11.6 COMBUSTIBLE GAS DETECTORS

These devices are widely used for monitoring flammable gases in factory atmospheres and can be applied to flue gas monitoring (Moseley and Tofield, 1987b). Their main use is as indicative devices for incomplete combustion rather than as quantitative analysers. They are often combined with zirconia or electrochemical cells for the determination of combustion efficiency.

The gases are passed over a heated ceramic bead (a 'pellistor') coated with a catalyst. Surface combustion of any organic compound in the gas stream takes place, resulting in the temperature of the bead rising. The temperature change is monitored using a thermocouple embedded in the bead or by using the electrical resistance of a platinum wire in the bead as a resistance thermometer. Comparison with an identical uncatalysed bead also exposed to the gas stream compensates for any other temperature fluctuations. The system is non-specific and the response depends on the heat of combustion of the individual species present. It will respond to CO and H_2 as well as organics. Clearly the method only works if there is excess oxygen present and some systems will blend auxiliary air with the flue gas to ensure that the sensor continues to respond even under fuel-rich combustion conditions. This is not a problem when such devices are used for flammable gas monitoring in air. In the occupational safety context calibration would be carried out with methane or propane and the scale relates to a percentage of the Lower Explosion Limit (LEL). In flue gas analysis for organics, a typical range would be 0–5% LEL for methane. Alternatively, the sensor can be optimised for response to CO with a range of 0–1000 ppm, for example. The resolutions quoted for the sensors in different applications are from 0.01% to as low as 0.001% (100 and 10 ppm, respectively).

11.7 MASS SPECTROMETRY

A comprehensive discussion of the applications of mass spectrometry to gas analysis is beyond the scope of this chapter. The high capital cost of mass spectrometry systems has inhibited their use in situations where only one or two components of a flue gas are to be monitored. Nonetheless, they have found wide application in the analysis of chemical process gas streams where there is a requirement for monitoring several different gaseous components at a number of different locations.

With computer control and automatic stream switching a very powerful multiple gas analyser is obtained.

The gas is introduced through a fine hole in the sample line into the spectrometer at very low pressure and is ionized by an electron beam, producing positively charged molecules and molecular fragments derived from the parent molecules. These are separated on the basis of their mass-to-charge ratio (m/z) using either a quadrupole mass filter or magnetic sector analyser. In a quadrupole analyser the ions travel along a path down the centre of a cylinder containing four parallel rods. An electrical potential of a specific strength and frequency is applied alternately to opposite pairs of rods. Only those ions of one specific mass-to-charge ratio will pass right through the fluctuating electric field to the detector at the other end. The electric field strength can be set so that one particular ion is detected, or scanned so that all ions in a wide range of mass numbers are detected in turn. In a magnetic sector analyser the ion beam passes through a uniform magnetic field between the poles of a magnet and is deflected. The radius of curvature of the path depends on the mass-to-charge ratio and the ion current is measured at a series of Faraday collectors where they are electrically neutralized and the ion current measured. The output of each collector corresponds to different mass numbers. The instrument can be set to detect specific mass numbers (as many as 16 in some instruments) or the mass spectrum can be scanned by varying the magnetic field. The mass range of 2–120 a.m.u. or even up to 200 a.m.u. may be covered.

Many gases may be detected using the parent molecular ion, but this may not be applicable to other cases where there is overlap or interference from more than one species; examples would include CO, N_2 and C_2H_4 on mass number 28, CO_2 and NO_2 on mass number 46, or molecules which are isomers such as the *n*- and *iso*-butenes on mass number 56. In such cases it is necessary to make use of the fragmentation patterns or to rely on peaks in the spectrum at mass numbers corresponding to isotopic species, e.g. carbon contains 1% ^{13}C as well as ^{12}C, oxygen 0.2% of ^{18}O as well as ^{16}O, sulphur contains 4% of ^{34}S as well as ^{32}S, etc.

Mass spectrometers can give very stable output and can measure very wide concentration ranges (from 100% down to ppm levels). Measurements on the selected set of components can be repeated at intervals of a few seconds. Given the increasing requirements for multi-component gas analysis of emissions to the atmosphere it is likely that this powerful technique will be more widely used in future. For example, some chemical plants may find it possible to combine their requirements for process gas and emissions monitoring in one instrument using the stream switching technique.

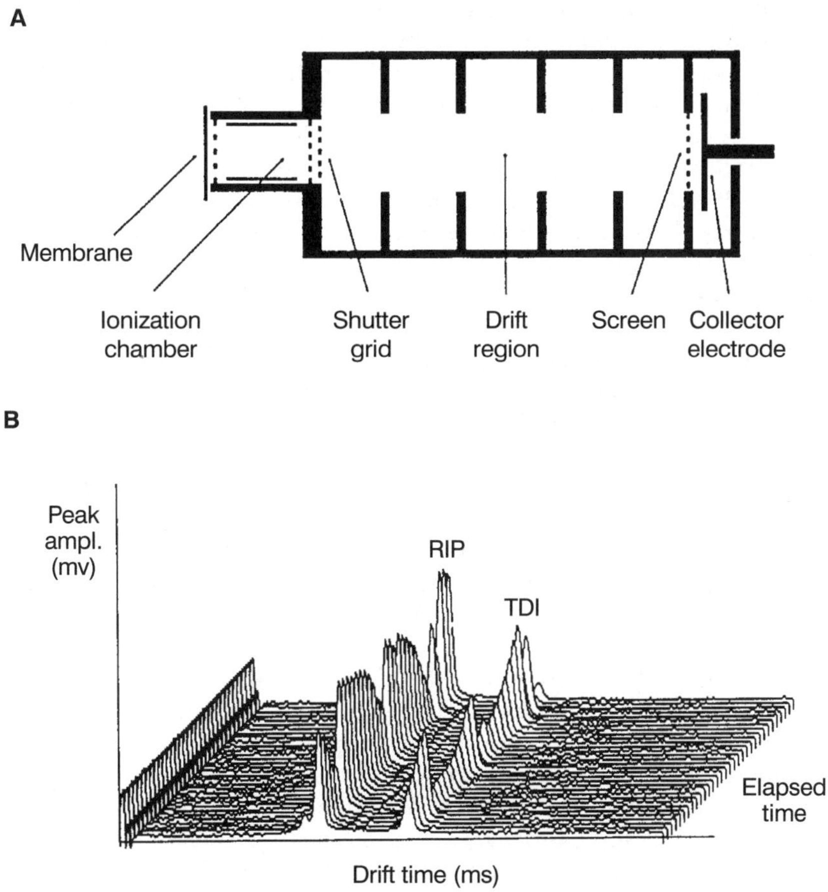

Fig. 11.10 (a) Schematic diagram of an ion mobility spectrometer. (b) Response as a function of drift time. Each trace represents the output from sequential sample gas pulses allowing the variation of concentration with elapsed time to be seen – in this case the gas is toluene di-isocyanate (TDI). RIP, reactant ions formed from air and water molecules. (Courtesy Graseby Andersen.)

11.8 ION MOBILITY SPECTROMETRY

Ion mobility spectrometry was developed originally as a monitor for the presence of chemical warfare agents. Large numbers of the hand-held portable versions are now in use with applications including detectors for explosives, drugs and hazardous vapours such as toluene di-isocyanate and hydrazines at sub-ppm levels. The application to flue gas monitoring has only appeared since about 1990 (Brokenshire and Pay, 1989).

The principle of the device is rather like a time-of-flight mass spectrometer operating at atmospheric pressure. The electrical mobility of an ion in a gas depends on its mass and the fundamental ion–molecule interactions which occur as the ion is transported through the gas under the influence of an electrical potential. Different ions have different mobilities and will take different times to traverse a given linear distance. The sample gas is introduced to an ionization section where a radioactive β source (^{63}Ni) converts some of the molecules to negative ions (Fig. 11.10). An electrical shutter consisting of a grid assembly, which can be maintained at different potentials, is operated so that a short pulse (e.g. 0.2 ms) of ions is introduced into the drift section. There the ions are transported under a potential gradient of the order of 250 V/cm taking a few 10s of milliseconds or less to drift to a collecting electrode where the ion current is measured. The current as a function of time since the opening of the shutter is the ion mobility spectrum. Scans can be repeated many times per second and the signals averaged. The current at the specific drift time for the ion of interest is proportional to the gas concentration in the sample. Background peaks in the spectrum relate to reactant ions or ion molecule clusters produced from air and water molecules, e.g. positive ions $[H^+.(N_2)_x.(H_2O)_y)]$ or negative ions $[(O_2)_n.(N_2)_x.(H_2O)_y]$.

For flue gas applications the instrument is tuned to a specific component. As yet, it is not possible to derive a multicomponent analysis from the mobility spectrum, unlike the mass spectrometers described in the previous section, nor is the selectivity as good. Although, as already indicated, the instrument can be very sensitive it does not have as wide a dynamic range as some other techniques. The high sensitivity does mean that it can be applied either to normal extractive or to dilution sampling systems. The main applications are for polar or electronegative gases such as NH_3, HCl, HF, Cl_2 and ClO_2 which are readily converted to negative ions. These are generally difficult gases to monitor by other techniques and here ion mobility spectrometry may offer a competitive alternative in the market.

REFERENCES

Brokenshire, J. and Pay, N. (1989) Ion mobility spectrometry: a promising technology for environmental measurement. *International Laboratory*, October.

Kocache, R. (1986) The measurement of oxygen in gas mixtures. *Journal of Physical and Electronic Scientific Instruments*, **19**, 401–12.

Moseley, P.T. and Tofield, B.C. (eds) (1987a) *Solid State Gas Sensing*, Chapter 1, Adam Hilger, Bristol.

Moseley, P.T. and Tofield, B.C. (eds) (1987b) *Solid State Gas Sensing*, Chapters 2 and 3, Adam Hilger, Bristol.

Moseley, P.T., Norris, J.O.W. and Williams, D.E. (eds) (1991a) *Techniques and Mechanisms in Gas Sensing*, Chapter 1, Adam Hilger, Bristol.

Moseley, P.T., Norris, J.O.W. and Williams, D.E. (eds) (1991b) *Techniques and Mechanisms in Gas Sensing*, Chapter 6, Adam Hilger, Bristol.
Moseley, P.T., Norris, J.O.W. and Williams, D.E. (eds) (1991c) *Techniques and Mechanisms in Gas Sensing*, Chapter 8, Adam Hilger, Bristol.
Moseley, P.T., Norris, J.O.W. and Williams, D.E. (eds) (1991d) *Techniques and Mechanisms in Gas Sensing*, Chapters 3 and 4, Adam Hilger, Bristol.

12

Calibration of continuous gaseous emission measuring systems

David S. Walker

12.1 INTRODUCTION

The calibration of continuous emission measuring systems (CEMS) is primarily concerned with determining the accuracy of the measuring system under investigation. Some operators of CEMS consider calibration to comprise simple zero and span checks plus adjustment, which involves setting the zero and span responses of the CEMS to the known values of standard reference gas mixtures (SRGM). Adjustment often follows from a zero and span check and the deviation or degree of adjustment required is sometimes used as a measure of accuracy. However, such rudimentary procedures cannot stand alone and qualify as valid calibration because the accuracy assessment is insufficient.

Various standard definitions of calibration are available (including the comprehensive definition given in British Standard BS 5233) and are a useful starting point when considering calibration procedures.

Calibration is arguably the most important component of the quality control of CEMS, and should be designed to satisfy the requisite standard definition and in particular meet the following objectives:

- to estimate the accuracy (precision and bias) of the measuring system;
- to satisfy regulatory requirements – including frequency of calibration and quality of reference standards;
- to employ appropriate standard procedures; and
- to realize the potential performance of the continuous emission measuring system.

Only a comprehensive and technically sound calibration procedure can achieve these objectives. The remainder of this chapter describes the main calibration procedures for CEMS and discusses their attributes and compatibility with the measuring system.

12.2 SAMPLING AND ACCURACY OF CEMS

When calibrating a CEMS the complete system must first be assessed on-site to check if the CEMS is sampling representatively. If this is not carried out, the accuracy of the CEMS cannot be determined and hence the main objective of calibration cannot be met. Restricting calibration to the analyser component is not a valid method because any bias due to unrepresentative sampling, sampling losses or transformation of sample is effectively ignored.

Checks for the representativeness of sampling include ensuring that *in situ* systems are installed at suitable sampling positions according to a recognized standard method. For extractive systems, sampling at a single point, it may be necessary to take measurements at various points on the cross-section of the flue to determine whether the CEMS sampling point provides sample with an average value of the flue gas concentration.

Isokinetic sampling is normally required in order to sample particulate matter representatively. However, when sampling gaseous emissions, isokinetic sampling is not usually needed for CEMS except perhaps when sampling downstream of wet scrubbers where water droplets could be present. A better alternative is to avoid such sampling positions if at all possible. Furthermore, to reduce the chance of reaction and adsorption on the particulate matter retained by line-filters, it may be necessary to adjust the sampling conditions to minimise the collection of particulate matter.

If the CEMS is shown to be sampling unrepresentatively there is little point in continuing the calibration until the measuring system is modified, perhaps by installing a network of sampling points or re-locating the CEMS at a suitable sampling position.

12.2.1 Estimating accuracy

The accuracy of a measuring system comprises components of random and systematic error (i.e. precision and bias). Carrying out a series of measurements reduces the impact of random error and tends to isolate the systematic error. However, when calibrating CEMS it is preferable to quantify both the precision and bias.

Gaseous emissions change their characteristics with time hence the repeatability (precision) of the CEMS can only be determined using SRGM which unfortunately cannot emulate the exact properties of flue gas. The precision can be expressed in terms of standard deviation (or coefficient of variation).

When estimating bias the whole of the CEMS must be considered unless the individual bias of a CEMS's component is being assessed. Current methods which are used for quantifying bias include: determining the relative error (which is described later in the chapter); statistical treatment of the calibration graph to estimate the limits of error; and a procedure based on standard deviation.

For statistical analysis, linear or quadratic regression should be used depending on whichever best describes the analyses function. A review of calibration statistics was given recently (Miller, 1991). The method using standard deviations was described in a recent dual BSI/ISO standard (BS 6069; 1992).

12.3 TYPES OF CEMS AND AVAILABLE CALIBRATION METHODS

There are essentially two types of CEMS: (i) those employing extractive sampling; and (ii) non-extractive or *in situ* systems employing sampling within the stack. The extractive CEMS can be sub-divided into direct sampling systems and dilution systems, which employ dilution probes to dilute the stack gas within the probe or in a conditioning system outside the stack. Similarly, the non-extractive systems can be divided into path monitors which sample across the stack using an optical beam, or point monitors which sample and analyse within the stack at a point or very short path. The main components of extractive and non-extractive gaseous emission measuring systems are illustrated in block diagram in Fig. 12.1.

Some CEMS are described by the manufacturers as self-calibrating and have internal reference standards inside the analyser which facilitate automatic zero and span checks. Although this facility can be a useful measure for monitoring the zero or span drift most, if not all, regulatory authorities insist on external calibration methods also being used.

There are essentially three external calibration methods available: (i) methods employing standard reference gas mixtures; (ii) methods using

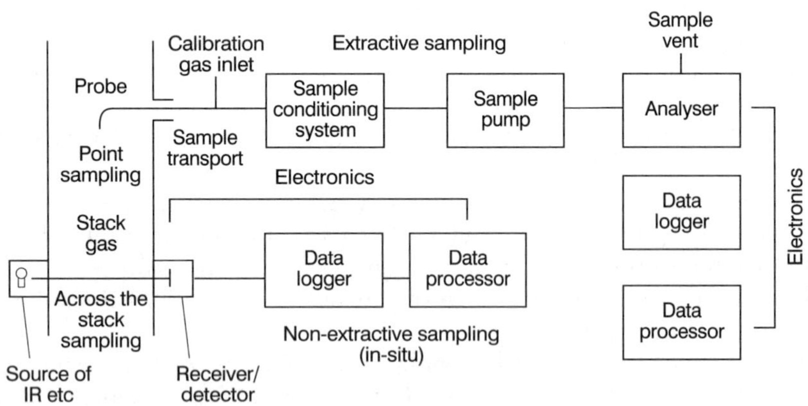

Fig. 12.1 Gaseous emission measuring systems.

comparison with standard manual reference test methods; and (iii) calibration methods using comparison with a previously calibrated instrumental system (verified CEMS).

Occasionally, a particular calibration method cannot be used for a specific type of CEMS. For example, it is usually impractical to calibrate a path monitor using gas mixtures. Selection of the calibration method depends on the type of CEMS and its application. In some countries, CEMS used for regulatory monitoring have to comply with the calibration method specified by the regulatory authority. However, in the UK the regulatory authority will usually accept any method tailored to the CEMS and the plant conditions, provided that good calibration is demonstrated.

With the exception of the previously mentioned dual BSI/ISO standard (BS 6069, 1992), there are no British Standard methods for calibrating CEMS. Some other national standards exist which address calibration, relevant examples include a Canadian standard (CAN/CSA-Z223.2-M86, 1986) and a German Verein Deutscher Ingenieure Guideline (VDI 3950, 1991). The US Environmental Protection Agency (USEPA) also has well-defined calibration methods (US EPA 600/9-005, 1984). International standards have been published on evaluating the performance standards of CEMS (ISO 8158, 1985) and the calibration of SO_2 CEMS (BS 6069, 1992), and others are in preparation including a standard method for calibrating nitrogen oxides emission measuring systems (ISO/DIS 10849, 1994). International methods published by the European Standards Committee (CEN) will probably be adopted in future European Union (EU) Directives. In the absence of CEN Standards suitable ISO Standards can be adopted by CEN and hence may also appear in future EU Directives.

12.4 CALIBRATION WITH STANDARD REFERENCE GAS MIXTURES

All external methods for calibrating CEMS rely on standard reference gas mixtures (SRGM) in some part of the method, particularly when checking the zero and span drift. However, some methods rely exclusively on comparison with SRGM and are discussed in this section. These methods essentially consist of zero and span checks together with the preparation of calibration graphs following a multi-point spanning. Methods employing gases are the most popular and probably the most likely to be carried out badly, mainly because of insufficient attention being given to obtaining and maintaining good quality standard reference gas mixtures, assessing losses of sample, or determining the effect of interferent gases on the measurements.

A discussion on calibration with SRGMs should begin with matters which affect the quality of the SRGM, such as the specification, preparation, analysis, stability, traceability, handling, etc. However, these items are not dealt with here, but they are discussed in detail in Chapter 13.

Calibration with SRGM has the advantage of convenience and lower cost, particularly in situations where sample losses and the presence of interferents are known to be insignificant and can be discounted. However, difficulties occur when interference errors or sample losses are encountered. The method cannot then be used, unless it is possible to compensate for these errors or to make accurate corrections, otherwise the accuracy of the CEMS cannot be determined. Unfortunately, estimating these errors is time-consuming and expensive and must be carried out on the CEMS on a site specific basis. Furthermore, major changes to plant operation – for example, a change in fuel or process – may require a reassessment.

Unless the calibration task specifically calls for the CEMS to be set up and adjusted before the calibration, no adjustments should be made. Although adjustments and routine operational checks form part of the quality control they are distinct from the calibration. All adjustments and checks should be made in accordance with the manufacturer's instructions. Care should be taken when adjusting the span and zero to ensure adjustment of one does not simultaneously affect the other.

12.4.1 Preparation of calibration graph

A calibration graph of the CEMS response versus the concentration of the SRGM is prepared by multi-point spanning, which involves introducing SRGM into the sample probe inlet (or as far upstream as possible from the sample conditioning unit) at the same flow rate used to sample the flue gas. Provided that insignificant sample loss is demonstrated, the introduction of the SRGM can be made directly into the analyser. A minimum of four concentration levels of SRGM at approximately 20, 40, 60 and 80% of the range should be used for linear CEMS and a minimum of 10 levels for covering the range of non-linear CEMS.

The calibration graph is normally prepared when: (i) a CEMS is first commissioned on-site; (ii) after repair; (iii) after a period of inoperation; (iv) at intervals specified by the regulatory authority; or (v) when the zero or span drift is found to be out of control.

Multi-point spanning is also required for non-extractive across-the-stack CEMS. However, it is not possible to use a dynamic procedure, and gas cells certified by the manufacturer are required.

When a multi-point span shows deviation from the previous calibration graph greater than the specified limits, data collected during this interval should be rejected.

When preparing the calibration graph several options are available, including manually drawing a line or curve to fit the data. However, a statistical treatment is usually preferred using linear or quadratic regression which are the methods generally adopted by German (VDI 3950, 1991) and US (EPA 600/9-005; 1984) methods.

12.4.2 Determination of analyser linearity

The response of most analysers is linear and is usually achieved via an electronic linearizing circuit. If faults occur in this circuit, substantial measurement errors can occur. Linearity is determined via multi-point spanning of the analyser only, to check if the calibration points deviate from the calibration line by more than the specified values.

12.4.3 Determination of sampling losses

Although extractive CEMS can experience loss of sample (or sample transformation) in the sampling and conditioning system, they have the advantage that the sample can be conditioned to meet the requirements of the analyser. Conditioning can involve many processes, including filtration, drying, temperature and pressure control, dilution, etc. Unfortunately, extractive sampling and conditioning can occasionally lead to serious loss of the sample. Therefore, when calibrating a CEMS it is necessary to quantify the loss of sample. In extreme instances the sampling and conditioning system may need to be modified to achieve an acceptable level of loss.

The loss of sample is assessed by alternately introducing SRGM directly into the analyser and probe inlet and comparing the responses. It is necessary to ensure that the temperature, pressure, humidity and flow rate of the SRGM entering the analyser is identical during alternate measurements. The sample loss can be expressed as follows:

$$\% \text{ loss of sample} = \frac{\text{AR} - \text{MSR}}{\text{AR}} \times 100$$

where AR refers to the analyser response and MSR to the measuring system response, i.e. to sample injected at the probe inlet.

12.4.4 Quantifying interference

All CEMS can experience cross-sensitivity or interference from other gases, including water vapour, which can cause major errors in the measurements. One solution is to remove the interferent. This is relatively easy for moisture, but cannot easily be achieved for most gases. Hence, when calibrating a CEMS, an investigation is required of the effect of other gases present in the sample on the CEMS response.

For extractive CEMS, interference can be assessed off-site or on-site using SRGM of gases suspected of being interferents. The effect of interferents on non-extractive CEMS cannot normally be achieved on-site using SRGM. Operators usually have to rely on the manufacturer's information on interferents, which is usually based on laboratory tests.

Consequently, when selecting cross-stack monitors, operators should ensure that the monitor has been shown to be sufficiently insensitive to the potential interferents present in the stack gas. Aspects which should be investigated include:

- the response of each potential interferent gas at the maximum concentration encountered in the stack;
- the relationship between the concentration of the interferent gas and the analyser response; and
- the synergistic effect of interferent gases when present together.

The method used for assessing interference in Germany is recommended (VDI 3950, 1991). The procedure comprises introducing zero gas downstream of the sample gas cooler/dryer and noting the response AR_O, followed by introducing the interferent gas and noting any change in response of the analyser AR_I. The interference (I, in %) is expressed relative to the full-scale deflection of the measuring range (MR) of the analyser according to the formula:

$$I = \frac{AR_O - AR_I}{MR} \times 100$$

When the interferent effect of water vapour is required, dry zero gas is introduced to the analyser via the sample gas cooler/dryer before measuring the response. The zero gas is then moistened to a dew-point of about 25°C and the operation repeated.

Correction to the calibration graph is made for a total interference error of ≤4%. Interference errors >4% require investigation and possible modification of the conditioning system or analyser.

An assessment of potential interferents should only be required during calibration following the installation of the CEMS or when either the performance characteristics of the CEMS is changed. For example, when a new analyser with a different principle of detection is installed, or the composition of the stack gas is changed substantially due to a change in the plant or process conditions.

12.4.5 Monitoring zero and span drift

Substantial drift in the zero or span usually indicates a change in the calibration graph. Therefore, checks on the zero and span are an important part of the calibration procedure and are required on a frequent basis. The frequency of zero and span checks should ideally depend on the characteristics of the particular CEMS which is being calibrated. However, CEMS drift characteristics can vary greatly and therefore it is better to standardize the frequency. Once every 24 hours has been adopted by many organizations, including the USEPA.

It is also necessary to consider a short-term frequency, in case short-term fluctuations greater than the 24-hour drift occur. The USEPA has adopted once every 2 hours as the frequency for checking short-term drift. Checks on short-term drift are usually only carried out periodically, initially to characterize the analyser and subsequently during full calibrations, or when special investigations are being carried out. When a CEMS is switched on after installation, repair, or inoperation, it should be allowed to stabilize before any drift determinations are made.

For extractive CEMS, the zero drift is determined by introducing air or nitrogen (free of the components to be measured) into the analyser under the same conditions of flow, temperature, etc. as the sample gas. After allowing sufficient time to purge the analyser, the response is noted. The span drift is determined similarly usually at one concentration corresponding to approximately the average value of expected values for the plant.

The zero and span drift should be recorded on control charts. Deteriorating drift or out-of-control drift can then be readily identified. The control limits are normally specified by the regulatory authority. It is essential therefore when choosing a CEMS that the manufacturer's specification for drift can meet the control limits.

A different procedure is required for non-extractive CEMS. The zero drift is usually checked by providing a zero condition in the light path by various means, for example inserting a tube to isolate the flue gas from the light path. Another means, mentioned in USEPA test procedures (Pate, 1983), is to insert three or more calibrated gas cells and compute the zero point from the upscale measurements. Similarly, the span should be checked against a gas cell certified to be functionally equivalent to its stated span concentration.

The previously described calibration procedures comprise the core components of a calibration method. Although there are other variables which can affect the CEMS performance, they can usually be ignored for practical calibration purposes. For example, the effect of changes in atmospheric temperature and pressure, voltage or vibration, on the response could be assessed, but are normally only required when the suitability of the CEMS for a particular application is being considered. However, for a full calibration it is advisable also to determine the response time and lower detection limit to check if the CEMS is functioning correctly.

12.4.6 Determination of response time

The response time is measured by passing SRGM into the sample probe inlet several times, and noting the average time interval between injection of the SRGM and the analyser response reaching 90% of the concentration of the SRGM.

12.4.7 Determination of lower detection limit

The lower detection limit is determined in a similar manner to the response time, except that zero gas is used and the response noted as quickly as possible to minimize the effect of zero drift. A minimum of 30 determinations should be carried out. Assuming a confidence level of 95%, the lower detection limit X (in mg/m^3) is given by the formula:

$$\underline{X} = X_O + 2\,S_{XO}$$

where X_O is the mean of the blank readings (in mg/m^3) and S_{XO} is the standard deviation of the blank readings in (mg/m^3). Full details, for measuring the response time and lower detection limit, are given in an ISO Standard (BS 6069, 1992).

Sometimes the design of the CEMS includes particular features which if not calibrated could have a substantial effect on the accuracy of the CEMS. For example, CEMS which utilize dilution probes and CEMS which have converters to convert nitrogen dioxide (NO_2) to nitric oxide (NO), and hence allow the measurement of NO_x (and NO_2 by difference).

12.4.8 Monitoring the dilution ratio of dilution probes

The dilution ratio is normally taken into account during multi-spanning when the various SRGM are passed through the dilution system. However, if it is suspected that the dilution ratio is not remaining constant during sampling, it should be checked – and monitored if necessary – by carrying out simultaneous monitoring of an undiluted component. For example, carbon dioxide can be monitored, via the dilution probe and in the undiluted form, to provide a real-time check on the dilution ratio.

12.4.9 Determination of oxides of nitrogen converter efficiency

The oxides of nitrogen (NO_x) converter efficiency can be measured using either static or dynamic methods to oxidize nitric oxide (NO). However, the static method, involving the use of a mixture of nitric oxide and oxygen in an inert plastic bag, is not recommended, because of the possible loss of sample within the bag.

The dynamic method depends on the rapid oxidation of nitric oxide with ozone relative to the very slow oxidation with oxygen. A simplified flow diagram of a test system for measuring the efficiency of NO_x converters is shown in Fig. 12.2. The method essentially comprises the following steps:

1. Zero and span the analyser.
2. Prepare a mixture of NO and nitrogen dioxide (NO_2) in the test system, by ozonizing a mixture of the NO span gas and oxygen.

3. Measure the NO content of the NO/NO_2 mixture with the analyser in NO mode (i.e. by-passing the converter).
4. Measure the NO_x content of the NO/NO_2 mixture with the analyser in the NO_x mode (i.e. via the converter).
5. Turn off the ozonizer in the test system, and measure the NO_x content of the NO and oxygen mixture with the analyser in the NO_x mode (i.e. via the converter).

The efficiency can be expressed as the ratio of the NO measured in the NO/NO_2 mixture, after passing through the analyser's converter, to the NO converted to NO_2 by the ozonizer in the test system.

When using this method, it is important to check that the analyser is not subject to interference from ozone. The detailed procedure for determining the efficiency of a NO_x converter is usually given in the manufacturer's instructions and is described in USEPA methods. The procedure should also be available shortly in an ISO standard method.

An alternative simple dynamic method for checking a converter's efficiency involves the use of SRGM containing NO_2 in synthetic air. However, if this method is used, the stability of the NO_2 SRGM mixture must be maintained and its concentration known with high accuracy.

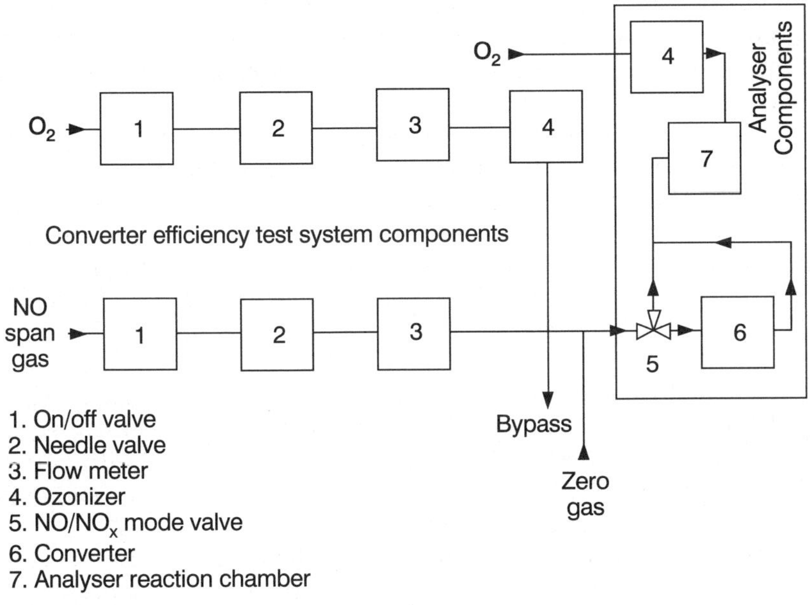

Fig. 12.2 Simplified flow diagram of converter efficiency test system.

12.4.10 Calibration with standard reference solutions

An analogous method of calibration is the use of Standard Reference Solutions (SRS) instead of SRGM for certain types of CEMS. For example, CEMS which operate by sampling the flue gas through a heated probe, followed by absorption into a transfer liquid before continuous analysis by an ion-selective electrode, often use SRS. However, some of these measuring systems also allow the use of SRGM for calibration, if it is preferred.

Since the accuracy of the calibration graph depends on the concentration of the SRS, it is important that the solutions are accurately prepared and used within their known lifetime. It is usually advisable to use freshly prepared SRS. When calibrating with SRS the manufacturer's instructions should be followed. However, similar to the calibration procedure with SRGM, a calibration graph should be prepared and regular checks carried out on the zero and span drift. It should be noted that CEMS of this type are also subject to sample losses and interferences. It may therefore in addition be prudent to calibrate by comparison with a manual SRTM.

12.5 CALIBRATION USING STANDARD REFERENCE TEST METHODS

An alternative method for calibrating CEMS is to compare the response of the CEMS with the results of manual measurements carried out simultaneously. This calibration depends on the manual measuring method providing the true value of the measurand, or more correctly, the conventional true value since the true value can never be known exactly. Therefore, it is essential that the manual measuring method selected is suitable for the measuring task.

Normally only a national or international standard reference test method is likely to suffice. However, even these methods may be found inappropriate in certain circumstances. For example, the standard manual method may be subject to interference from specified components. When using this method of calibration it is necessary to ensure that: (i) a suitable national or international standard method is used; (ii) the effect of interferents or sample losses are minimal; and (iii) the manual sampling is representative.

Calibration by comparison with a manual reference measuring method depends on a completely independent source of measurements. Hence, it is essential that a completely separate set of apparatus comprising probe, sampling line and train is used. For example, it is not permissible to utilize a sample of gas from the CEMS probe.

Calibration with a standard manual method has the advantage of checking the complete CEMS. If any part of the CEMS is seriously malfunctioning, out of control, or poorly adjusted, the comparison measurements should provide poor agreement.

This method also has the advantage that it does not depend exclusively on SRGM. Consequently, if sub-standard SRGM has inadvertently been used for adjusting, and zero and spanning the CEMS, this method will provide a check. Another advantage is the option for calculating the relative accuracy of the CEMS. Disadvantages are mainly those of cost – for example, the need for accredited chemical analytical facilities (in the UK, NAMAS accredited).

The procedure for this calibration method comprises carrying out a manual measurement strictly according to the standard, while recording the CEMS response. A sample of flue gas should be extracted for a set period, preferably specified in the method, but of sufficient duration to allow the collection of a representative sample. The CEMS should be set to simultaneously log and average the continuous measurements over this period. Ideally, a series of comparison measurements should be carried out over a range of pollutant concentrations.

Day-to-day calibration of the CEMS still requires checks on zero and span drift, occasional checks on linearity, and the calibration curve supplied by the manufacturer. Additional calibration checks for example on NO_2 to NO converters may also be required. Recommendations on the suitability of some standard manual methods for the calibration of CEMS are given in Table 12.1.

12.5.1 Determination of relative accuracy of CEMS

The relative accuracy of the CEMS under calibration can be expressed in terms of error and is determined by comparing the average value measured by the CEMS, during the period of calibration, with the value obtained from simultaneous measurement with the SRTM. Relative accuracy is used widely in the US and is calculated as follows (De Wies *et al.*, 1989):

$$\text{Relative error} = \frac{\text{Measured Value} - \text{Conventional True Value}}{\text{Conventional True Value}} \times 100$$

where the Measured Value is the average of the logged results from the CEMS and the Conventional True Value is the result from the manual standard reference method.

The concept of relative error is also compatible with British Standards and a definition of relative error is given in BS 5233.

Until fairly recently, the concept of calibrating CEMS by means of comparison with other measurement methods was restricted to comparison with standard manual methods. However, with the improvements in performance and reliability of automated methods, calibration of CEMS by comparison with automated methods has been increasingly practised in the UK and the US and is now recognised in ISO methods.

Table 12.1 Standard manual methods for CEMS calibration

Component measured	Method	Remarks
Sulphur dioxide	BS 6069: Section 4.1:1990 ISO 7934:1989	Dual issued standard
Nitrogen oxides	No national or international standard method recommended	USEPA Method 7C* May be suitable for CEMS at power stations or nitric acid works
Hydrogen chloride	VDI 3480 Part 1: 1984 or AFNOR X43-309: 1991 or USEPA Method 26*	A CEN standard method is currently being prepared
Sulphuric acid (Mist/SO_3)	BS 1756: Part 4: 1977 USEPA Method 8*	Also measures sulphur dioxide
Carbon monoxide	No national or international method recommended	BS 1756: Part 4:1977 Iodine pentoxide method is seldom used
Hydrogen fluoride	No national or international method recommended	A modification of USEPA Method 13* to determine gaseous fraction may be suitable

* USEPA Reference Methods in Code of Federal Regulations, 40 CFR 60 and 40 CFR 61

12.6 CALIBRATION USING VERIFIED CEMS

The calibration of a CEMS by comparing its response with that from another transportable or mobile verified CEMS, is similar to the method employing comparison with a manual standard reference test method. However, it has the additional advantage of real-time analyses; hence, any major deficiencies in the calibration of the CEMS under investigation is quickly perceived.

To qualify as verified, the reference CEMS's performance should be determined with regard to interference, sampling losses, sample transformation, etc. with a view to eliminating or correcting for systematic errors.

When using this method two completely separate systems should be used, i.e. separate probes and conditioning systems (if two extractive systems are being compared), and each CEMS's analyser should preferably have detectors utilizing different principles of operation.

As with the calibration method using a SRTM, both the CEMS being calibrated and the verified CEMS must sample the same gas stream. If extractive systems are being used the probe inlets would normally be adjacent within about 100 mm. However, if this is not possible and one probe is downstream from the other or far apart, it is necessary to establish that both systems are sampling the same unchanged parcel of flue gas.

Calibration with a verified CEMS is expensive and is normally only carried out infrequently, for example during plant commissioning, during specified regulatory calibrations, or when the plant or its mode of operation is changed substantially. Regular routine calibration operations, for example, zero and span drift assessment and occasional checks on linearity are integral parts of this calibration method, and are also required to ensure that the CEMS operates within control limits. This method also allows the determination of relative error.

12.6.1 Dual CEMS calibration system

A further development of comparing the CEMS being calibrated with one verified CEMS, is to make the comparison with two verified CEMS. This has the advantage of each of the reference CEMS continuously providing a quality control check on the other. Out-of-control conditions in either of the reference CEMS are easily identified and real-time analysis means that early correction can be achieved. Results from two verified CEMS operating on-site are given in Table 12.2 to demonstrate this approach.

The results for each measuring period ranging from 30 to 120 minutes were averaged and the standard deviation and relative accuracies calculated. During periods 1 to 7, good agreement was obtained. However, during periods 8 to 12, although the two CEMS indicated a similar precision, much poorer agreement was observed for the averages. A consistent but much poorer relative accuracy was also noted indicating serious bias in one of the CEMS. It is not obvious from the results which CEMS is at fault. Subsequent investigation confirmed that CEMS A had malfunctioned and on correction good agreement was again observed for the readings from each CEMS (see measuring period 13, after which measurements ceased).

12.6.2 Comparison of national and international methods

To conclude the chapter, the UK, FRG, USA and ISO procedures for calibrating CEMS are compared (Table 12.3). Most, if not all, of the ISO standard methods should in due course be adopted in the UK by BSI. Some of the ISO methods may also be adopted by CEN and hence could be included in future European Union Directives on topics involving continuous monitoring, for example, the Large Combustion Plant Directive. However, it seems very unlikely that ISO methods will supersede USEPA methods in the USA.

Table 12.2 Dual CEMS calibration carried out on-site

Measuring Period	Nitric oxide concentration (ppm)		Relative accuracy of A(%)
	CEMS A	CEMS B	
1	461	491	–6.1
2	439	468	–6.2
3	430	454	–5.3
4	497	482	3.1
5	483	491	–1.6
6	414	422	–1.9
7	483	491	–1.6
Average	458	471	Range –6.2 to 3.1
Calculated standard deviation	31.3	26.0	
Relative standard deviation (%)	6.8	5.5	
8	557	458	21.6
9	529	435	21.6
10	563	463	21.6
11	507	417	21.6
12	563	463	21.6
Average	544	447	–
Calculated standard deviation	24.9	20.5	
Relative standard deviation (%)	4.6	4.6	
13	503	518	–2.9

Table 12.3 Comparison of national and international CEMS calibration procedures

	UK	FRG	USA	ISO
Methods	1. Standard Reference Gas Mixtures, includes: regular zero and span checks.	1. Functional Test includes: checks for gas leakage, calibration function (curve or factor), interference, calibration drift, etc. 2. Analytical Test, comparison of CEMS response with approval manual method.	1. Relative Accuracy Test Audit and Relative Accuracy Audit, comparison with USEPA manual reference method. 2. Calibration Gas Audit and Calibration Drift Assessment, using cylinder of standard reference gas mixtures.	1. Three methods permitted: Standard Reference Gas Mixtures, checks on interference, sampling losses, etc. and preparation of calibration curve. 2. Comparison with ISO approved manual method. 3. Comparison with instrumental measuring system (validated according to ISO procedures).
Regulatory authority	Environment Agency (EA) for England and Wales.	Federal Environmental Agency (Umweltbundesamt UBA).	US Environmental Protection Agency.	–
Mandated in regulations	No	Yes	Yes	No (however, could be adopted by CEN and then included in future European Union Directives).
Choice of methods	Yes – approval by EA is required.	No	Limited – can now use instrumental methods for comparison.	Yes – recommendations given on choice of methods.
National or International Standard	No, but national manual standards are specified as reference methods.	No, supported by Verein Deutscher Ingenieure Guidelines.	No, USEPA methods	Yes, ISO 7935
Further information available from	EA and process guidance notes published by HM Stationery Office.	UBA, VDI guidelines and Air Pollution Control Manual (1988).	USEPA, Code of Federal Regulations, and Pate (1983)	ISO 7935:1992(E).

REFERENCES

Air Pollution Control Manual of Continuous Emission Monitoring. Regulations and Procedures for Emission Measurements, 2nd Revised Edition. Published by the German Federal Minister for the Environment Nature Conservation and Nuclear Safety, Berlin, 1988.

BS 5233 (1986) *Glossary of Terms Used in Metrology*. British Standard (Incorporating BS 2643).

BS 6069 (1992) *Stationary Source Emissions – Determination of the Mass Concentration of Sulphur Dioxide – Performance Characteristics of Automated Measuring Methods*. Section 4.4:1993/ISO 7935:1992 (E).

CAN/CSA-Z223.2-M86 (1986) *Method for the Continuous Measurement of Oxygen, Carbon Dioxide, Carbon Monoxide, Sulphur Dioxide and Oxides of Nitrogen in Enclosed Combustion Flue Gas Streams*. National Standard of Canada.

De Wies, W.G. *et al.* (1989) Recommended QA/QC for Municipal Waste Combustion Testing. Presented at the 82nd Annual Meeting & Exhibition, Anaheim, California, June, 1989. *Journal of the Air and Waste Association*.

ISO/DIS 10849 (1994) *Stationary Source Emissions – Determination of the Mass Concentration of Nitrogen Oxides – Performance Characteristics and Calibration of Automated Measuring Systems*. Draft International Standard.

ISO 8158 (1985) *Evaluation of the Performance Characteristics of Gas Analysers*. International Standard (E).

Miller, J.N. (1991). Basic statistical methods for analytical chemistry Part 2. Calibration and regression methods: a review. *Analyst*, **116**, 3–14.

Pate, W.J. (1983) Regulations and Resource File of Continuous Monitoring Information. Research Triangle Park, NC: USEPA, EPA-340/1-83-008.

US EPA Reference Methods in Code of Federal Regulations, 40 CFR 60 and 40 CRF 61.

US EPA-600/9-005 (1984) *Quality Assurance Handbook for Air Pollution Measuring Systems: Vol I Principles* and *Vol III Stationary Source Specific Methods*. Research Triangle Park, NC.

VDI 3950 (1991) *Calibration of Automatic Emission Measuring Equipment*. Verein Deutscher Ingenieure Draft Guideline Part 1.

13

Calibration gases

John Scawin

13.1 INTRODUCTION

Calibration gases are fundamental to the validity of industrial air pollution monitoring. Since most monitoring equipment used in this field is comparative, it is necessary to use a calibration gas to calibrate any analysers or sensors on a regular basis. Hence, it is necessary to select the most appropriate calibration gas standard for these determinations bearing in mind any further requirements of accuracy, traceability and stability that may be required by the pollution control authorities.

13.2 CALIBRATION GAS SELECTION

Calibration gas selection is dictated by various requirements. Firstly, as a basic rule, the gas mixture should duplicate the sample gas as much as possible. The concentration of the minor component should be as near as possible to that of the sample. If two gas mixtures are available then they should bracket the component concentration in the sample. The composition of the balance gas should be the same as that of the sample. Only with these precautions is it possible to eliminate matrix effects between analysers.

Traceability requirements will be specified by the authority that defines the measurement procedures. Stability requirements can only be determined historically. If a gas mixture is unstable then it will only be discovered over a period of time and then by comparison with another gas standard. Recommendations as to stability can very often only be obtained from a proven gas manufacturer who determines this property by the comparison of several gas mixtures over many years.

13.3 CALIBRATION GAS MIXTURE PREPARATION

For ultimate accuracy, calibration gas mixtures must be prepared from first principles, that is, by mixing known amounts of high-purity constituents. The different methods available for this purpose can be summarized as follows:

- Gravimetric methods
- Static volumetric methods
- Dynamic volumetric methods

The purity requirement of gases used in these preparations must be emphasized. All gases used must be carefully analysed and allowances made for any impurity present in the final mixtures. Impurities that affect other gases in the mixtures or have a cross-response on analytical sensors should be removed by purification before use. In high-purity mixtures this is now recognized as the largest error contribution present.

13.3.1 Gravimetric mixtures

This is the most accurate method for the preparation of calibration gas mixtures (ISO 6142). The basic requirement for the preparation of a calibration gas mixture by mass is a robust analytical balance (Fig. 13.1), the sensitivity of which to the mass of the minor component in the gas mixture is the main limiting factor on accuracy. A gas with a high molecular weight will therefore give greater accuracy than low-molecular weight gases.

In low concentration mixtures, it is possible to overcome the inaccuracy introduced by the low initial weight of the minor component in the mixture by the preparation of the same gas mixture with this component at a higher concentration. After through mixing, the weighing operation is repeated using an aliquot of this mixture with more balance gas. This dilution step is a useful technique which can be repeated three or four times, to achieve low concentrations of minor components in gas standards. Extra care must be taken with the purity of the balance gases at all stages and adsorption effects within the cylinders. Dilution steps should be minimized by the selection of a balance with very high sensitivity.

In addition, while the principle of making a mixture of gases by mass is simple enough certain precautions must be taken if the results are to be of any value. For example,

- great care is needed in the preparation of the weighing vessel;
- complete dryness and evacuation are absolutely essential;
- equilibrium conditions must be maintained before and after weighing;
- weighing should be carried out in a controlled humidity and temperature environment; and
- the effects of buoyancy must be allowed for either by using a counterpoise giving a similar displacement as the cylinder or by calculation.

For further information, the reader is referred to ISO 6142.

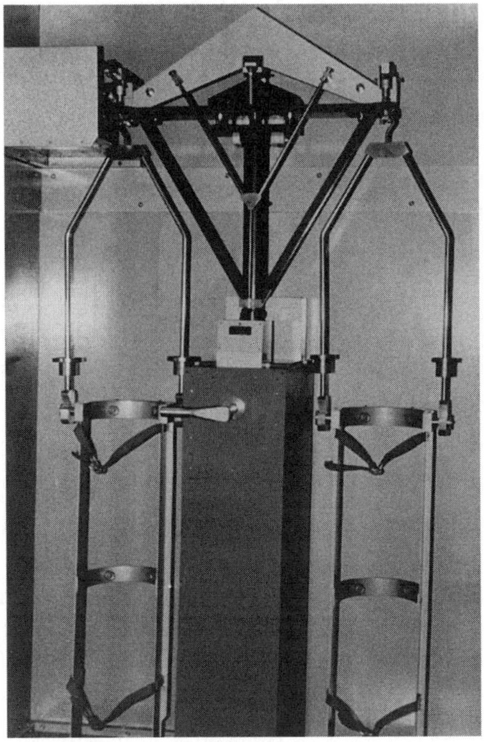

Fig. 13.1 A 100 kg balance. (Courtesy Reverifications Ltd, Aylesbury.)

13.4 STATIC VOLUMETRIC METHODS

The most direct method for the preparation of gas mixtures is to mix volumes of gases which have been accurately defined at ambient temperature (ISO 6144). This method has been developed into the partial pressure technique used by most gas manufacturers to prepare high-pressure mixtures.

The components of the mixture and the balance gas are injected successively into a cylinder presumed to be of constant volume and which has previously been evacuated. The pressure is measured after each injection. The final concentration of the calibration gas is then equal to the ratio of the pressure change caused by the introduction of the calibration component to the total pressure of the gas mixture. This is expressed in terms of a pressure ratio. The conversion of this ratio into a volume concentration can be calculated by various methods, the most common being those of Dalton, Amagat and Kay.

Temperature has a major effect on the accuracy of the result. As each component is introduced into the mixture, any increase in temperature must be taken into consideration. The compressibility, Z, of a gas is another major factor. In addition, since the mutual diffusion of components is generally not rapid enough to homogenize the mixture during filling, one observes a piston effect on one component when the next is introduced. When homogenization finally takes place the mixture has a lower coefficient, but with the same mass and there is an increase in pressure. This pressure increase can be as much as a few percent and would lead to a considerable error if the cylinder were not finally topped-up after waiting for homogenization.

Due to the inaccuracies that arise as a result of their partial pressure preparation these mixtures are named (concentration defined) by their analysis against a more accurate comparison standard. If the comparison standards used incorporate reference material from a national body then these mixtures become traceable by definition. This procedure will be further discussed under traceability later in this chapter

13.4.1 Syringe injection methods

A measured volume of gas is injected through a rubber septum into a calibrated glass container, e.g. 1 litre. A further volume of complementary gas (50 ml) is injected with a syringe and the contents mixed by alternatively extracting and re-injecting the 50-ml portion. After several passes to ensure mixing is complete, 50 ml of sample can be withdrawn for analysis.

To maximize accuracy, all volumes should be checked by filling with water and weighing. Temperature variations will also have extreme effects; all gases must be brought to laboratory temperature before use and care must be taken to ensure that they are not warmed during injection.

This method has been extended to very large volumes (20–200 1itres), with pressurized additions of complementary gas and with a motorized fan to achieve efficient mixing to provide low ppb concentrations of atmospheric contaminants in air for calibration purposes.

13.5 DYNAMIC VOLUMETRIC METHODS

This method which can be used to provide a flow of calibration gas depends on the blending of two or more gases, the flows of which must be accurately controlled and measured. The accuracy of the blending system together with the control and measurement of the flow determines the total accuracy of the method.

13.5.1 Mixing pumps

Two gases are aspirated together by a series of pumps at ambient temperature and pressure. Lower concentrations may be obtained by cascading the mixture from one pump to another. In some pumps the gases are confined over oil, but in later models the oil has been eliminated. Different concentrations are produced by selection of varying gear ratios.

The assembly is calibrated by comparison of the gas mixtures produced with traceable gas mixtures provided by a gas supplier or a National Institute.

13.5.2 Periodic injection

A small volume (0.2–0.3 ml) of gas is periodically injected into a known flow of balance gas. The rate of injection is controlled by a motor-driven injection device. Concentration can be varied by altering the injection rate and the complementary gas flow (ISO 6145).

13.5.3 Continuous injection

The minor component is injected continuously through a capillary tube, by means of a mechanically driven piston, into a steady flow of balance gas. This method is applicable to volatile liquids which vaporize into the complementary gas flow (ISO 6145).

13.5.4 Gas blenders

These are commercially available systems which blend two gases through a series of capillary tubes to provide a selected range of gas mixtures with a claimed accuracy of 0.1 to 1.0% relative. The latest models are computer-controlled to allow for compressibility differences between the component and the complementary gas mixture. As the gas mixture is varied the results from an associated analyser are scanned and printed out as a calibration curve

13.5.5 Mass flow controllers

These are blending systems operated by two or more mass flow controllers which combine to provide a flow of calibration gas. The gas flow is metered by a valve which is controlled electrically by variations from thermal conductivity sensors in the systems. Such systems have a claimed accuracy of ± 1.0 % and can be used on corrosive gases (BS 4559).

13.5.6 Permeation tubes

Permeation tubes consist essentially of a tube of porous material, usually Teflon, containing a readily liquefied gas. This permeates through the walls of the tube at a constant rate dependent upon temperature (ISO 6349). These devices are used to generate atmospheres of known concentration by placing them in a gas stream flowing at a fixed rate and constant temperature. Care must be taken to ensure that the tubes give a predictable permeation rate even after subjection to varying temperatures both in use and storage. The latest systems incorporate a gravimetric balance to weigh the permeation tube in the complementary gas flow and compute the final concentration.

13.5.7 Saturation methods

Gas flowing at a measured rate is saturated by passage through a saturator at known temperature (ISO 6145). This gas is then passed through a constant bath held at 5°C below that of the saturator. The concentration of the gas mixture can be calculated from the saturation values at the observed temperature. Such systems can be use to prepare moisture standards in gases (ISO 6147).

13.5.8 Choice of methods

All of the above methods and many others not described are available for the preparation of calibration gas mixtures. The final choice however, will depend on the accuracy required for the gas mixture and the properties of the components, i.e. their stability. The accuracy provided by each method is summarized in Fig. 13.2.

The data in Fig. 13.2. indicate that the gravimetric method at high pressure is the most accurate method for the preparation of calibration gas mixtures. However, if the gas mixture contains a corrosive or reactive component, then this method is unsuitable since the final product may decay in the container. In these circumstances it is necessary to change to a dynamic volumetric method and the user will be advised to adopt a permeation or gas blending technique to prepare a calibration gas mixture.

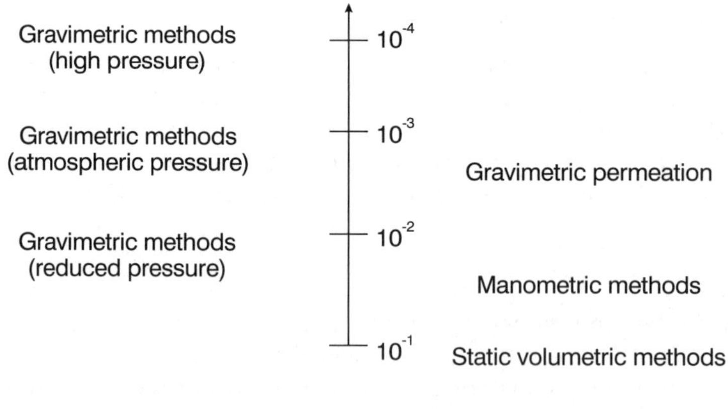

Fig. 13.2 Uncertainties of preparation methods for calibration gas mixtures.

Users should be aware that over and above the requirement for accuracy, the controlling authorities now require a traceability concept in the calibration and measurement procedures carried out in industrial air pollution monitoring.

13.6 TRACEABILITY

- **Traceability** is the property of the result of a measurement or the value of a standard whereby it can be related to stated references, usually national or international standards, through an unbroken chain of comparisons all having stated uncertainties (ISO 14111). This concept is often expressed by the adjective 'traceable', while the unbroken chain of comparisons is called a 'traceability chain'.
- **Accuracy** is the closeness of the agreement between a measurement result and the true value of the measurand (BIPM/IEC/OMIL/IFCC/IUPAC).
- **Trueness** is the closeness of agreement between the average value obtained from a large series of measurement results and the true value of the measurand (BIPM/IEC/OMIL/IFCC/IUPAC). The measure of trueness is usually expressed in terms of bias.
- **Precision** is the closeness of agreement between independent measurements obtained under prescribed conditions.
- **Uncertainty** is an estimate attached to a measurement result which characterizes the range of values within which the true value is asserted to lie.

13.7 TRACEABILITY OF CALIBRATION GAS MIXTURES

Most speciality gas manufacturers will supply traceable data with their calibration gas mixtures. Users should satisfy themselves that these data are available and are sufficient to comply with mandatory regulations. The following procedure by which a gas manufacturer provides traceable calibration gas standards is given by kind permission of Special Gases, BOC.

13.7.1 Gas purity

All gases used in the preparation of traceable gas standards must be analysed and checked for purity before use. Should any interfering contaminants be found present, then they must be removed by purification. To prevent any deterioration in subsequent use the cylinders and filling lines associated with this mixture preparation must be scrupulously cleaned and dried. If the gas mixture contains any reactive component, then any passivation treatment of the cylinders and equipment must be carried out before filling.

13.7.2 Primary standards

Gas manufacturer's primary standards (GMPS) are prepared by weight at high pressure. (see section 13.3.1) using an ultra-sensitive balance (Voland mode1 1115 CDN; balance load 10 kg ± 1 mg). Assessments of balance performance are carried out at 6-monthly intervals.

The weights used are certified class S, and they have been subsequently revalidated by the National Physical Laboratory (NPL) or a NAMAS-approved laboratory. To prevent any equilibrium changes during filling, the laboratory is controlled to a temperature of 21± 1°C, and a relative humidity of 55 ± 5%.

Starting with two pure gases, e.g. carbon monoxide and nitrogen, a series of GMPSs is prepared. This series of standards is called a parentage tree because lower-concentration mixtures are prepared by a double dilution weighing technique from higher concentrations. The validity of the parentage tree and the GMPSs within it can be carried out by comparison analysis of the low concentration standards. If these agree among themselves, then the higher standards are correct. This principle is demonstrated in Fig. 13.3.

13.7.3 Correlation

The GMPSs have been produced by weight and their accuracy is assessed at better than ±0.1% relative. In the process of correlation they are correlated with national (NPL) or international (NMI, NIST) standards and their traceability can be established.

The procedure that follows is laid down by the Environmental Protection Agency (EPA) USA, TP102 and agreed with the NAMAS authorities. In a specified range the GMPS are analysed in conjunction

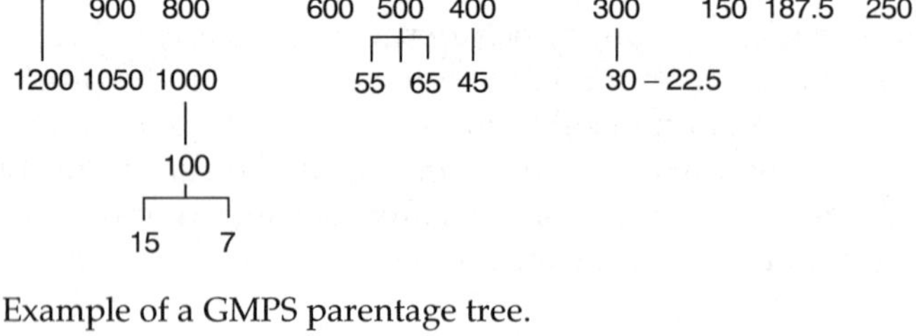

Fig. 13.3 Example of a GMPS parentage tree.

with one or more of the national standards. The analytical responses of the GMPSs and the national standards, which are weighted in the computerized programme, are prepared as a calibration curve. In the curve the molar concentrations derived from their analytical responses and their gravimetric preparation are compared. Provided that both results agree within ±0.5%, then the calibration is accepted as valid.

Since national and international standards are compared on an international basis, they are accepted as equivalent. On this basis their concentration result can be excepted as the 'true' value. This is the first step of the traceability chain which has determined the 'trueness' of the GMPSs. i.e. 0.5%. An example of a calibration is shown in Table 13.1.

Table 13.1 shows the international standards (column 1), the analytical response in digital volts (DVM) (column 2), and the respective weighting (column 3). Note that the national standards have weighting of 100. In columns 4 and 5 are shown the molar concentrations, column 5 containing results calculated from the analytical data. The percentage difference between columns 4 and 5 is shown in column 6; this difference must be <0.5% of a point in order for the calibration curve to be valid.

13.7.4 Calibration curve validity

Calibration curves are repeated over a period of time to allow for changes in detector sensitivity or where an individual gas standard is emptied and has to be replaced. Repeat curves are rejected if differences >0.5% occur from the computerized line, or any points of inflexion are noted. Provided that these criteria are accepted, then the new curve is declared valid, numbered and dated and incorporated in the laboratory records.

Table 13.1 Example of a calibration curve prepared for carbon monoxide standards

Cylinder no.	Analytical response (DVM)	Weighting	Cylinder concentration	Calculated concentration	Difference (%)
E 846	163	1	66.6005	66.8641	0.39
CAL 5141	235	100	97.1000	97.0933	0.00
E 720	241	1	99.4982	99.6331	0.13
E 992	359	1	149.8127	150.2634	0.30
E 966	435	1	183.7665	183.5942	–0.09
54022	696	1	302.6856	302.7969	0.03
53685	851	1	377.5559	377.3661	–0.05
E 703	955	1	429.3066	429.0961	–0.04
CAL 5570	1045	100	475.0000	475.0062	0.00
53710	1102	1	504.9720	504.6448	–0.06

13.7.5 Gas naming of traceable standards

Traceable standards are prepared as gas mixtures using pure gases following the volumetric method (see section 13.4). Where there is any possibility of instability due to the presence of a reactive component, then a passivation treatment is applied to the internal walls of the cylinder. The prepared mixtures are placed in the correlation laboratory and allowed to equilibrate with the controlled environment of 21°C and 55% relative humidity.

After 24 hours, the cylinders are analysed by direct comparison with the two nearest bracketing GMPSs taken from a valid calibration curve (ISO 14111). The analysis is repeated 7 days later and the two results compared. The results from both analyses must agree within 0.5% for the analysis to be accepted. This second analysis is carried out to improve the reproducibility of the process.

This gas-naming process is the second step of the operation and completes the traceability chain which started with the national (NPL) standards, continued with the GMPSs, and ended with traceable gas mixtures. Both steps can now be combined to give a traceability of:

GMPSs traceability	±0.5%
Traceable gas mixtures	0.5%
Total	±1.0%

Because of the derived relationship of <±1.0% to national standards, Special Gases, BOC, called these gas mixtures ALPHA standards.

Much information is available on each ALPHA standard and it is important to ensure that it is passed to any user who may require to satisfy the authorities with the traceability aspects of their analyses. This information is given on the certificates provided with each standard.

13.7.6 Certification

The certificates provided with each ALPHA standard will carry the following details:

- A statement that the standard is traceable to 1.0% to national standards.
- The manufacturing order number and cylinder number.
- Date of preparation.
- Two analytical results; one at day 2, and the second at day 9.
- The concentrations of the two bracketing standards with their calibration curve numbers.
- Cylinder pressure recommended stability period and analytical method used.
- Analytical method and units.
- Signature of approved person.

For audit surveillance, all of the above records and procedures are available for inspection purposes.

13.8 STABILITY

Despite the care that goes into gas mixture preparation, all the procedures described so far can fail completely if the gas mixture is unstable. It is therefore necessary for any gas manufacturer to take every precaution to ensure stability in their cylinders.

The stability of a gas mixture can only be determined on a historical basis, i.e. changes over a period of time. This means that the analyst must have an absolute standard available or further gas mixtures to use for comparison purposes. It is unlikely that two or more cylinders will deteriorate at the same rate and therefore one should be able to detect differences between them. This would be good evidence of instability.

Instability can take many forms, as shown diagrammatically in Fig. 13.4. This indicates some of the instability effects that can occur when reactive gases are blended in a steel cylinder with an inert balance gas. The concentration of such mixtures can vary with time, temperature or pressure. The nature of the instability is random and dependent on the particular condition of the individual cylinder involved. Very often, the

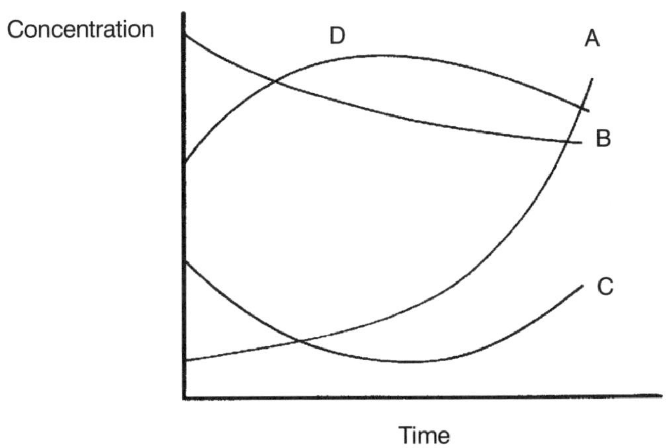

Fig. 13.4 Diagrammatic examples of instability of gas mixtures in cylinders. Curve A: desorption effect. An example of the increase in moisture content from a cylinder during use as the pressure falls. Curve B: catalytic reaction. An example of the reduction in carbon monoxide content in a carbon/manganese cylinder with time. Curve C: adsorption/desorption. An example of a low concentration of carbon dioxide mixture in a wet rusty cylinder as the pressure falls. Curve D: desorption/adsorption effect. An example of a low concentration of sulphur dioxide in a dry rusty cylinder with temperature.

exact mechanisms involved are still not fully understood (BIPM/IEC/IFCC/IUPAC; EPA Test 102/105; Rhoderick and Zielinski, 1988).

For the gas manufacturer producing calibration gas mixtures the first consideration is the material compatibility of the cylinder and valve. If there is any reaction between these and the gases in the cylinder then the final mixture will be unstable. Of particular importance are the components of the valve, many of which are plastic and highly unsuitable.

The next consideration is cylinder preparation. The internal surfaces of the cylinder must be clean, dry and without any trace of oil, grease or anything that may react with its future contents. During this process any heavy rust deposits should be removed.

Stability trials should be carried on various gas mixtures using different materials and treatments. These trials can be organized using the system shown in Fig. 13.5. This illustrates how four evacuated cylinders can be filled from a premixed container. The pressure from this single bulk container is allowed to equalize into the four cylinders in such a way that each cylinder receives the same charge. All cylinders, including the bulk container, are analysed against each other over a period of time until any significant results are found. By varying treatments and concentrations of gases it is possible to evaluate the best treatment for any reactive component.

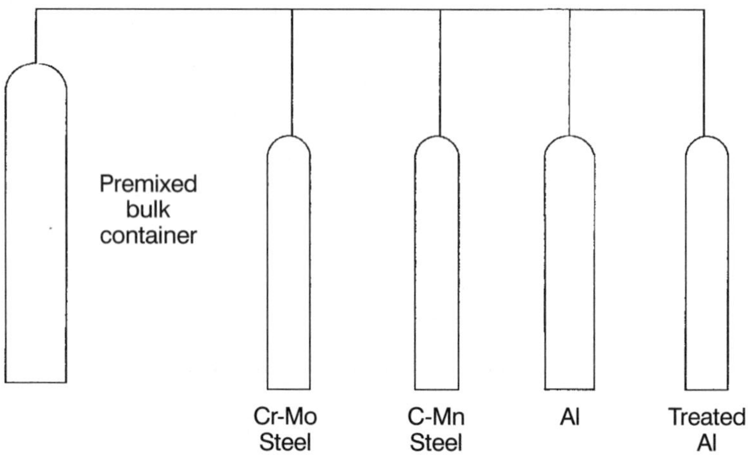

Fig. 13.5 Filling system for stability testing of different cylinder materials and varying internal treatments.

13.9 'SPECTRA SEAL'

Special Gases, BOC, have evaluated mixtures of reactive gases for many years and have perfected a proprietary process for the treatment of aluminium cylinders. This process is called 'Spectra Seal'. The basis of the process is to seal the minute pores found in the surface of the aluminium oxide layer that always coats metallic aluminium. Once sealed in this way, the adsorption and desorption effects are considerably reduced. For the first time, pressure and temperature variations have no effect on the concentration within the cylinder.

This treatment is so efficient that it is quite possible to prepare moisture standards between 1 to 500 ppm which are quite stable and have been evaluated as transfer standards by the NPL (McGivern, 1989). The same treatment is applied to ALPHA standards where they contain reactive gas components – a procedure which enables the company to claim a 5-year stability period on such gas mixtures.

REFERENCES

BIPM/IEC/OMIL/IFCC/IUPAC: *International Vocabulary of Basic and General Terms in Metrology (VPM)*, Second Edition 1993.

EPA Test Procedure TP 102/105 (1977) Calibration of gas mixtures/Gas naming. US Federal Register 85-075-23(a), Vol. 43, No. 24.

ISO 14111 (1996) Natural gas – Guidelines for traceability in analysis.

ISO 6142 (1981) Gas analysis – Preparation of calibration gas mixtures – Weighing methods.

ISO 6144 (1983) (BS 4559, Part 3) Gas analysis – Preparation of calibration gas mixtures – Static volumetric methods.

ISO 6145 Part 3 (1986) Periodic injections into a flowing gas stream (BS 4559, Section 5.3, 1987).

ISO 6145 Part 4 (1986) Continuous injection method (BS 4559, Section 5.4, 1987).

ISO 6147 (1979) Preparation of calibration gas mixtures – Saturation method (BS 4559, Part 6, 1981).

ISO 6349 (1979) Preparation of calibration gas mixtures – Permeation method (BS 4559, Part 7, 1981).

McGivern, W.H. (1989) Evaluation of gas phase moisture standards for use as calibration transfer devices. NPL report VOM89, September.

Rhoderick, G.C. and Zielinski, W.L. (1988) Preparation of multicomponent gas standards of toxic organic compounds in the low parts-per-billion range. *Analytical Chemistry*, **60**, 2454–60.

Wechter, S.G. and Kramer, N.P. (1975) 21st National Symposium, A.S.F.M.

14

Data logging

Richard Grant

14.1 EMISSION MONITORING REQUIREMENTS

Recording data from instrumentation is an important part of any monitoring system. With a growing environmental conscience, industrial plants can be fined or shut down for producing high emissions. Thus, emission monitoring systems have become critical. This chapter examines the different methods used to evaluate processes against legislation. Chart recorders, data loggers and Personal Computer-based systems are discussed.

In the UK, local authorities and Her Majesty's Inspectorate of Pollution (now incorporated into the Environment Agency) issue process guidance notes to industry. These detail what levels of pollutants that a particular plant is permitted to emit. Different plants and processes have different permissible emissions, which must be measured using different criteria. Legislation demands some special features of any system which attempts to prove or disprove compliance.

14.1.1 Normalization

Emission data must always be reported with reference to standard levels of temperature, oxygen, pressure and water vapour – referred to as normalization. This involves measuring these parameters and then converting the emission data. Some of these conversions (particularly oxygen) can have a dramatic effect on the emission levels.

Temperature: correction = Temperature (K)/273
Pressure: correction = Measured pressure (kPa)/101.3
Oxygen: correction $= \dfrac{20.9 - \text{Standard } O_2 \text{ level } (\%)}{20.9 - \text{Measured } O_2 \text{ level } (\%)}$

Standard oxygen levels vary according to fuel type and process, typical levels are 3% for oil or gas, 6% for solid fuel.

Water vapour: correction $= \dfrac{100}{100 - H_2O\ (\%)}$

Normalization is conducted because HMIP must compare like with like, and measurements can be diluted easily by adding excess air or water injection after the combustion process, deliberately or otherwise. Normalization corrects for any dilution, and refers densities to a standard volume at 0°C at 1 bar (101.3 kPa), referred to as a normal metre cubed (Nm^3).

14.1.2 Averaging

Pollutant emission levels are almost always stated over a particular averaging time – a rolling average. This may be thought of as a clockface of numbers – emission levels, the oldest number being replaced with the latest number at a regular interval. The rolling average is the average of all numbers on the clockface, and the averaging time is time for one complete revolution. This may vary from a few minutes to a few days. Also, operators are quite often allowed to exceed their emission levels for a percentage of time. This permits the occasional plant problem or operator error to be accepted, without a breach of the permissible emissions. There may be a requirement for several averages, each with its own different emission limit.

14.1.3 Typical guidance note

A typical emission limit table for industry is shown in Table 14.1. These guidance notes require the emission data to be referred to: 0°C, 6% oxygen (solid fuel), 101.3 kPa, and dry conditions.

Averaging times used for these notes vary from a calendar month to a 2-hour rolling average. Also, a limit is placed on the percentage of time for which all 48-hour rolling averages are allowed to exceed a level. There is also a requirement that any averages are not diluted by plant off-periods, any system should be able to record and indicate these periods.

Table 14.1 Typical emission limit table

	Calendar monthly average should not exceed	No more than 3% of all 48-h averages should exceed in a calendar year	The average over 2-h continuous operation should not exceed
Particulates	100	110	90
SO_2	2000	2200	
NO_x	650	715	

14.1.4 Data storage

Data are normally required to be stored for a period of 4 years, either as paper records from chart recorders, or as a computer data storage medium, which vary from floppy disks to re-writeable compact disks, etc.

14.1.5 Summary

It can be seen that an emission monitoring system must be capable of performing a high level of data processing to assess a plants emissions. There are two prime requisites for any data logging system which is to be used for monitoring pollutant emissions:

1. It must allow inspectors to assess a plants emissions with respect to legislation.
2. Operators must be given enough information to be able to run their plants without exceeding the emission limits.

Systems which attempt to meet these two requirements vary from simple chart recorders to data loggers and computer-based systems.

14.2 METHODS OF EMISSION MONITORING

14.2.1 Simple chart recorders

Simple analogue devices have been in existence for a long time, and consist of a moving strip/disk of paper pulled underneath a pen. A current or voltage is produced by the instrumentation and this is transmitted to the recorder, and drives the movement of the pen. The result is a strip/disk of paper with a pen trace(s) representing the emission levels. A simple system is shown in Fig. 14.1.

Although simple chart recorders are still widely used, they suffer from some major drawbacks:

- Annotation for time and other information is completed by hand. This is prone to operator error and may be unreliable and inaccurate.
- Only a limited number of channels can be accommodated.
- Time base (typically 20 mm/h) and scaling are fixed and cannot be altered after the trace is drawn.
- Examining data collected some time ago is not easy. A roll of chart paper may represent a month's worth of data and going back several months or years would not be particularly easy.
- Little information is provided to the operator with regard to normalization data: if a high emission level was recorded, why?, what was the oxygen, temperature, etc?

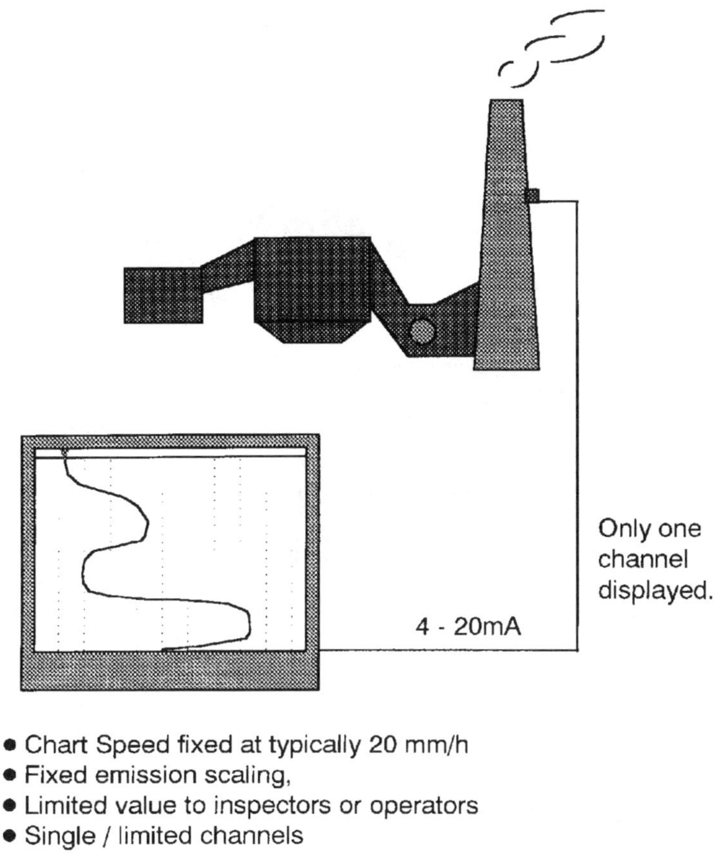

Fig. 14.1 Single-channel chart recorder.

- Only one average may be considered, and the instruments must be capable of providing this average.
- If the level exceeds the full scale of the recorder, the actual levels are lost.
- Should the recorder run out of paper or ink, all data are lost.

Such simple systems are easy to understand, but do not satisfy either of the requisites for an emission monitoring system.

14.2.2 Multi-channel recorders

Modern chart recorders can record the levels for many channels, and can offer automatic time annotation and calculations between channels – for normalization. An example of such a system is shown in Fig. 14.2.

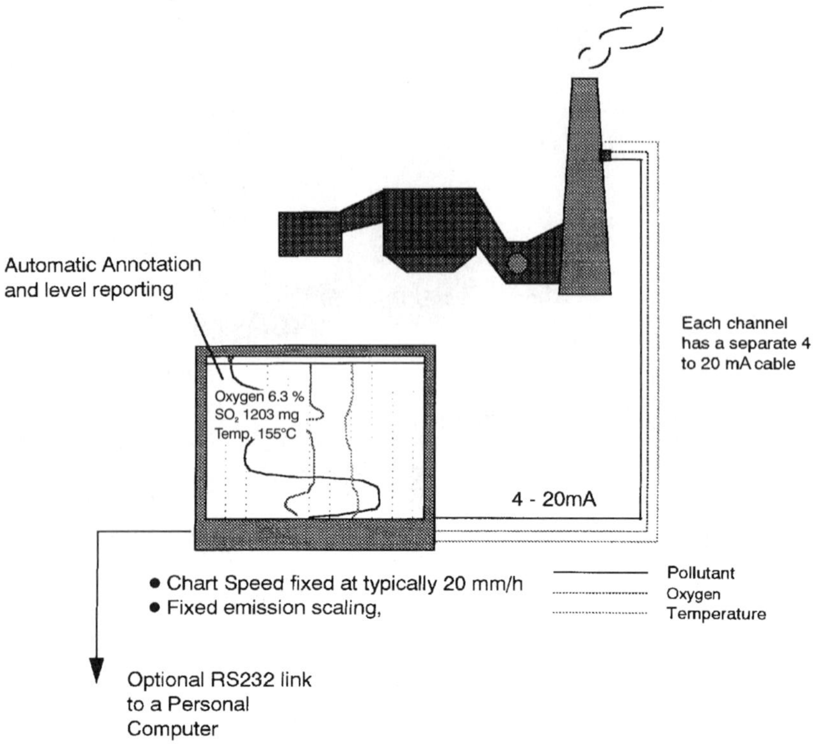

Fig. 14.2 Multi-channel chart recorder.

These recorders can provide multiple rolling averages, but the data still exist only as paper – although some recorders have a limited memory capacity and can download data onto a computer at regular intervals. This is a 'messy' technique to transfer the data into a computer and involves regular operator interference.

A modern, sophisticated chart recorder can meet the requirements of legislation, but still retains some of the drawbacks of a simple chart recorder. It is not so easy for inspectors to assess a plants emissions using the data recorded by them (rolls of paper), particularly with respect to complex legislation.

14.2.3 Data loggers

Small data loggers are available for simple, single-measurement applications. These are digital replacements of the single pen analogue chart recorder. Instead of being stored on paper, the data are stored within internal Random Access Memory and available for display, either graphically or numerically. Some instruments offer the capability to download the data onto a personal computer (PC), but this transfer is a clumsy technique. An example of a typical simple system is shown in Fig. 14.3.

Data processing is quite limited – typically no normalization or rolling averages. While such data provide an indication of how a plant is operating, it would be impossible for an operator to assess a plant's emissions, if the legislation was complex. Sophisticated data loggers are also available which are much more capable and are included in PC-based systems.

Until now, all of the systems examined lack sufficient processing capability to be able to process the emission data to a level required by most legislation. Although it is possible to transfer the data to a PC, multiple rolling averages are only available historically, not as they are building

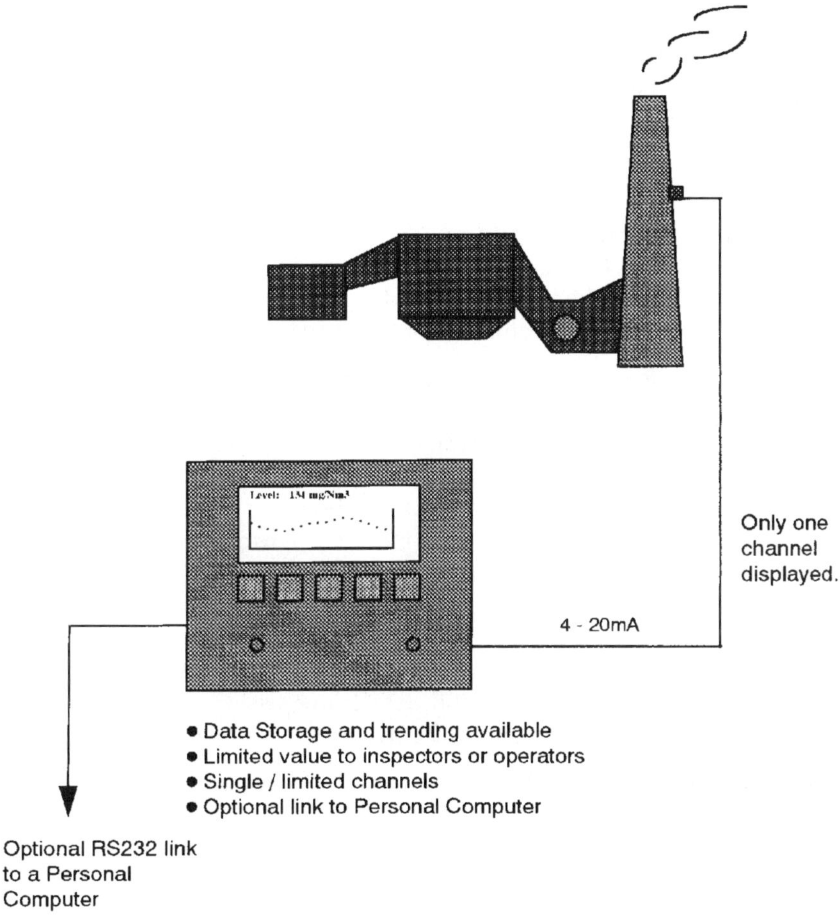

Fig. 14.3 Single-channel data logger.

up. They could be used by HMIP to examine whether a plant has been operating within emission limits, but not by plant operators attempting to control emission levels for different rolling averages, as they happen.

14.2.4 Personal computer-based systems

Personal computers have many features, most of which are of significant benefit to operators and inspectors with the demands of emission monitoring. These include:

- extremely fast processing power to create real-time rolling averages and normalized measurements;
- massive data storage both to store, and allow access to, many years of emission data;
- full colour display to show the emission data as required; and
- interface to commercial printers to print data as required.

Once the data have been transferred to a PC (in real-time) the possibilities are only limited by the program in operation.

14.2.5 Data conversion

Most instrumentation provide analogue signals which represent emission data; before a PC can be used for emission monitoring, these analogue signals must be converted into a digital format. This can be conducted in one or a combination of two ways:

1. By using analogue to digital (A/D) converters, which convert the analogue signals into numbers. Depending upon the type of A to D, this number is normally in the range 0 to 255 (8 bit) or 0 to 4095 (12 bit). These numbers are then transmitted to the PC on a standard interface in a standard format – normally RS232. Some of these systems can also be data-loggers, providing data storage, alarms, and real-time data transfer to the PC.
2. By using a networked system where the instrumentation transmits real numbers to an interface unit, which then transfers them to the PC in a similar fashion as above. This system requires microprocessor-based instruments and a digital data transmission. Instruments without microprocessors can use A/D converters as above.

The method of converting/transmitting the signals has an effect on the capability of the system.

Analogue data transmission/conversion (Fig. 14.4)

An analogue data transmission (normally 4 to 20 mA) requires the input to be scaled correctly. The A/D may be scaled say from 0 to 500 mg of pollutant, the 4 mA position relating to 0 mg, and 20 mA to 500 mg. The output of the measurement transducer must, of course, match this range. The range has an effect on accuracy and must be chosen carefully: too small a range means that important data will be lost at levels above the full-scale position; too large a range means that the resolution at lower levels will be poor.

These systems can meet the requirements of the most complex legislation. They do not, however, allow the emission monitoring system to be used as a diagnostic tool for instrument operation. It is not possible to send diagnostic data on the current loop, other than instrument failure where less than 4 mA is transmitted.

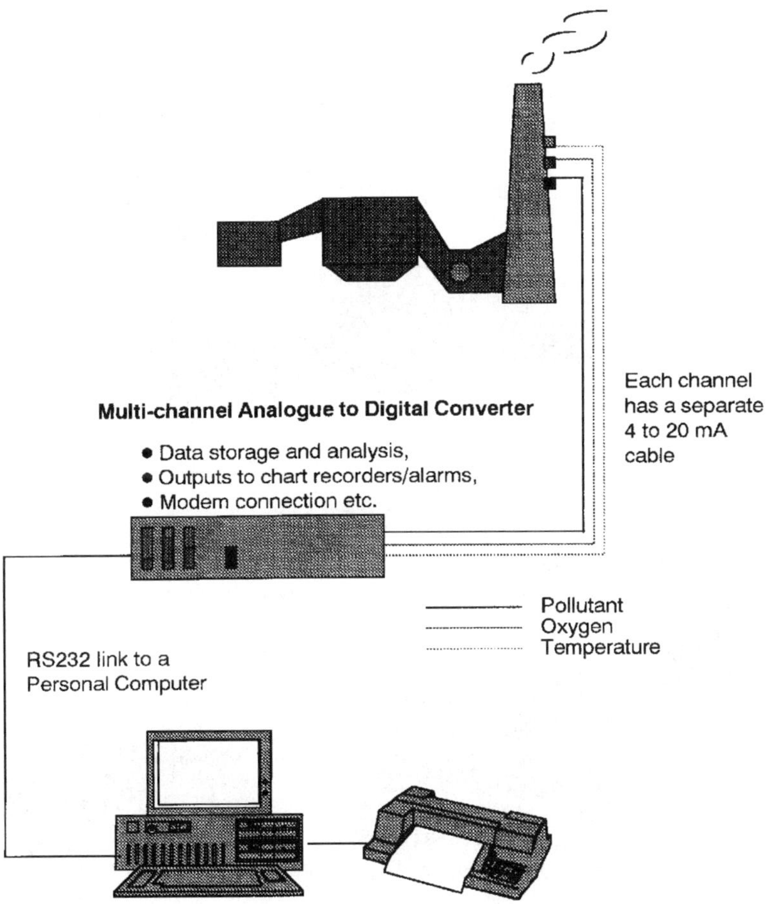

Fig. 14.4 PC system based on analogue to digital data conversion.

Digital data transmission (Fig. 14.5)

Using microprocessor-based instrumentation, the data are already in a digital format, at some stage in the measurement the analogue signals will have been converted to a digital format; this usually occurs at a low level, i.e. the instrument processes numbers, not analogue signals to calculate emission levels. The emission levels are transmitted to the PC via an interface. This digital transmission system can send much more information than the analogue system, e.g. instrument diagnostics, maintenance indications, several rolling averages, as well as raw and normalized data. This information can then be used to assess the performance of the instrumentation and predict maintenance requirements, etc.

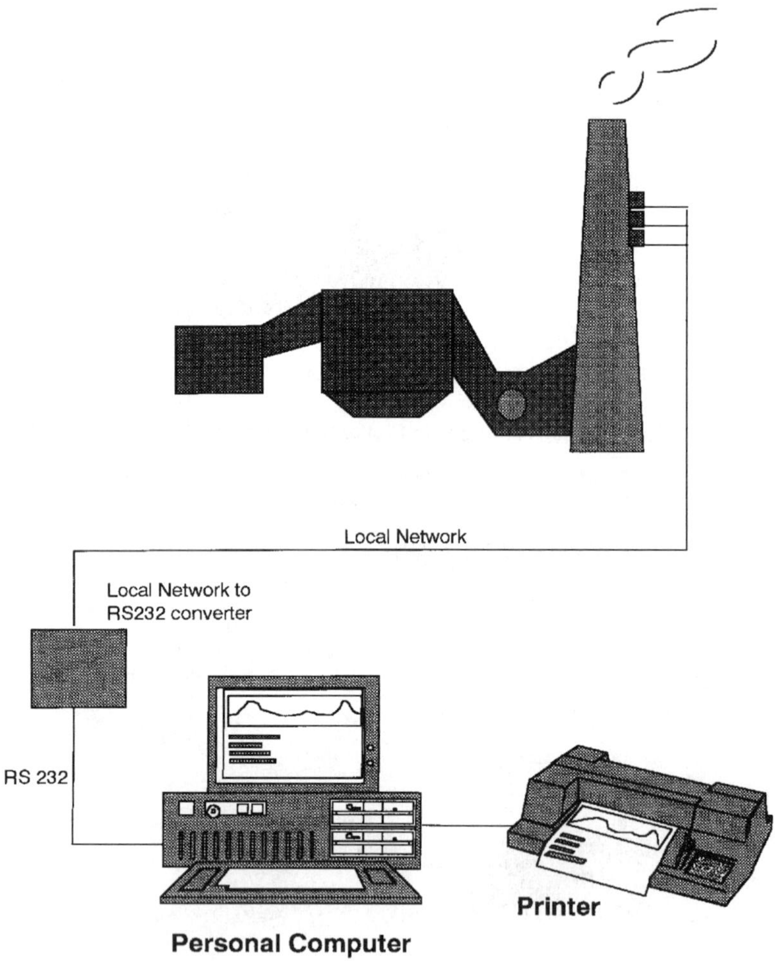

Fig. 14.5 PC system based on digital data transmission.

14.2.6 Software

Once the data exist within the PC in real-time, from either an analogue or digital source, the possibilities are only limited by the software in operation. Typical capabilities of a computer-based system are now examined.

Real-time data

Operators need to know what the emission levels are for several rolling averages in real-time. Figure 14.6 displays a bar graph representation of all measurements for one boiler. The data have been averaged and are displayed in terms of mg/Nm³, as required by legislation. Other averages and units are available for display.

Single channel trends

Data have been trended for two channels in Fig. 14.7. The pictures represent a 24-hour view of the data, with a point plotted every 10 minutes. Each data point has been stored labelled automatically for time. A cursor is available which can be positioned anywhere on the curve and the level at that particular time examined. A bar across the top of each trend on this particular system gives an indication of plant status and data validity.

Multi-channel trends

Figure 14.8 illustrates several channels plotted for the same time base for a single boiler (on a PC screen, each channel would be displayed in a

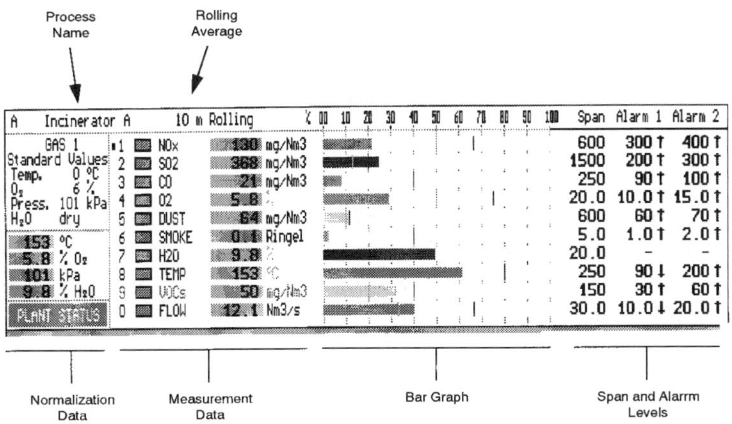

Fig. 14.6 PC display – typical real-time data.

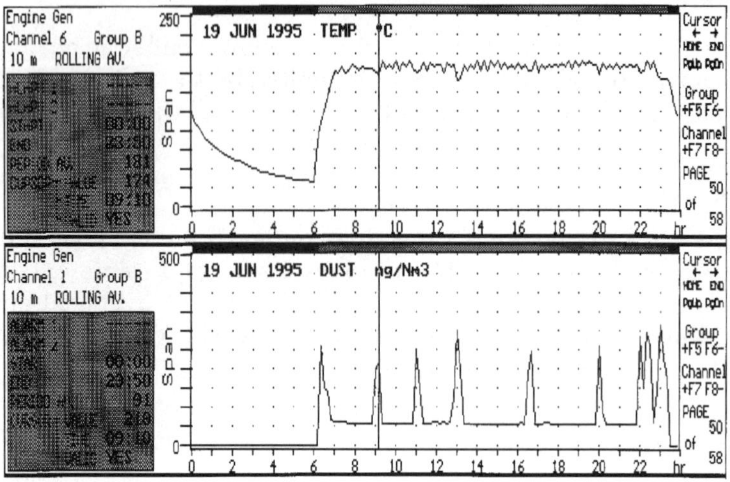

Fig. 14.7 PC display – single channel trends.

different colour). While this information is useful to plant operators to see how pollutants change with different plant conditions, it is of limited use to the inspector. A cursor is available to access the levels of all measurements at a particular time.

Excursions above alarm levels

A system can analyse the emission data for excursions above a definable level for a selected rolling average. Figure 14.9 shows an analysis of a system for a particular day. It also examines whether the plant has kept within permissible excursions within a defined period, a common requirement with many guidance notes. This analysis could be conducted for any rolling average over any period.

Average levels

Perhaps the first step for any examination of emission levels would be to produce a summary, where average and maximum data are examined over a selectable period, as illustrated in Fig. 14.10. This system also gives an indication of the validity of measurement and plant operating times.

Maxima analysis

It is useful for plant operators and inspectors to be able to analyse a long period of data to determine the periods of highest emission. This is possible of course by examining data for every measurement on a day-by-day

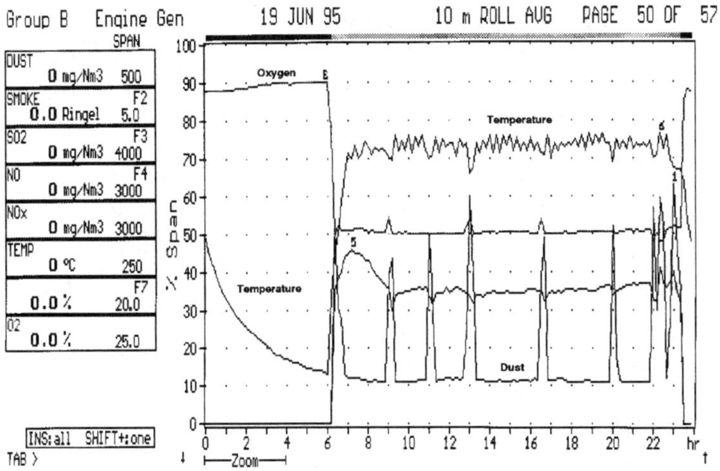

Fig. 14.8 PC display – multi-channel trends.

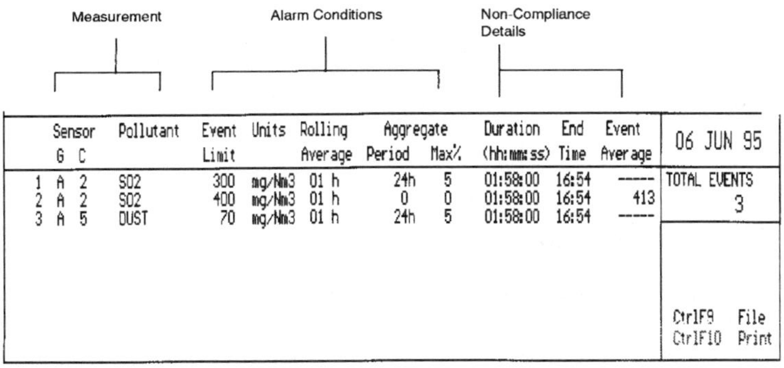

Fig. 14.9 PC display – excursions above alarm level analysis.

basis, but even with a computer this would be time-consuming business. Figure 14.11 is an analysis over 3 months for a system, and highlights the three highest peaks for each boiler and measurement, and the date and time at which they occurred.

Distribution of data

Figure 14.12 illustrates the percentage of time for which an emission measurement has been within specific bands – 20 bands are considered here. Using these and other analysis of the emission data, plant operators are able to run plants within emission limits, and inspectors can demonstrate compliance or non-compliance with legislation.

DAY Summary report for 19 JUN 1995.

Pollutant	Units	Average	Maximum	Tonne	Valid	Plant
B1 DUST	mg/Nm3	90.5	315.0	N/A	100.0	71.5
2 SMOKE	Ringel	0.42	1.36	N/A	100.0	-
3 SO2	mg/Nm3	570.9	872.0	N/A	100.0	-
4 NO	mg/Nm3	676.6	862.0	N/A	100.0	-
5 NOx	mg/Nm3	1089.1	1388.0	N/A	100.0	-
6 TEMP	°C	181.0	192.0	N/A	100.0	-
7 H2O	%	4.3	11.9	N/A	100.0	-
8 O2	%	12.7	13.7	N/A	100.0	-
C1 DUST	mg/Nm3	0.0	0.0	N/A	100.0	0.0
2 SMOKE	Ringel	0.00	0.00	N/A	100.0	-
3 SO2	mg/Nm3	0.0	0.0	N/A	100.0	-
4 NO	mg/Nm3	0.0	0.0	N/A	100.0	-
5 NOx	mg/Nm3	0.0	0.0	N/A	100.0	-
6 TEMP	°C	0.0	0.0	N/A	100.0	-
7 H2O	%	0.0	0.0	N/A	100.0	-
8 O2	%	0.0	0.0	N/A	100.0	-
D1 DUST	mg/Nm3	8.8	30.0	N/A	100.0	100.0
2 SMOKE	Ringel	0.02	0.07	N/A	100.0	-
3 SO2	mg/Nm3	1559.2	1806.0	N/A	100.0	-
4 NO	mg/Nm3	359.8	410.0	N/A	100.0	-
5 NOx	mg/Nm3	578.9	660.0	N/A	100.0	-
6 TEMP	°C	143.5	148.0	N/A	100.0	-
7 H2O	%	7.5	7.5	N/A	100.0	-
8 O2	%	6.2	8.8	N/A	100.0	-

Fig. 14.10 PC display – average level report.

14.2.7 Real-time inspection

So far we have only discussed how a plant may be examined by inspectors historically, and not how it is operating now. PC-based systems and sophisticated data loggers can be attached to a MODEM, this allows the information displayed and stored to be transferred by telephone, and would allow inspectors to inspect plants as they are operating.

These systems could also offer remote maintenance watchdogs to ensure that monitoring equipment operates at the highest availability.

14.3 SUMMARY

Chart recorders and basic systems do not satisfy the needs for logging emission information. Data analysis is either difficult or impossible and is limited to crude evaluations from rolls of paper.

Systems which transmit data to personal computers in real-time perhaps offer the best solutions to the problems posed by emission monitoring. Their data processing, storage, display and printing capabilities provide everything that is required now, plus the ability to cope with any future requirements by incorporating only changes in software.

QUARTER Maxima Report Starting MAY 1995, using 10 m rolling average data.
Peaks of only more than 110% above each daily average examined.

Group	Channel	Date	Time	Level
Boiler 6	DUST	02/06	05:00	1757
Boiler 6	DUST	03/06	21:30	1450
Boiler 6	DUST	02/06	01:00	1393
Boiler 6	SMOKE	03/06	21:30	1.8
Boiler 6	SMOKE	02/06	05:00	1.8
Boiler 6	SMOKE	05/06	20:20	1.4
Boiler 6	SO2	02/06	05:10	7995
Boiler 6	SO2	03/06	01:00	7694
Boiler 6	SO2	03/06	05:30	7628
Boiler 6	NOx	02/06	05:10	813
Boiler 6	NOx	02/06	05:30	785
Boiler 6	NOx	02/06	11:10	782
Engine Gen	DUST	29/05	15:30	5076
Engine Gen	DUST	26/05	15:00	5049
Engine Gen	DUST	30/05	09:20	4206
Engine Gen	SMOKE	26/05	18:20	4.4
Engine Gen	SMOKE	13/06	22:30	3.0
Engine Gen	SMOKE	26/05	15:40	3.0
Engine Gen	SO2	26/05	20:50	2660
Engine Gen	SO2	26/05	20:20	1132
Engine Gen	SO2	08/06	00:00	1019
Engine Gen	NOx	26/05	21:40	4554
Engine Gen	NOx	26/05	14:50	1810
Engine Gen	NOx	27/05	01:20	1626
Boiler 5	DUST	No Peaks Found		
Boiler 5	SMOKE	No Peaks Found		

Fig. 14.11 PC display – emission maxima report.

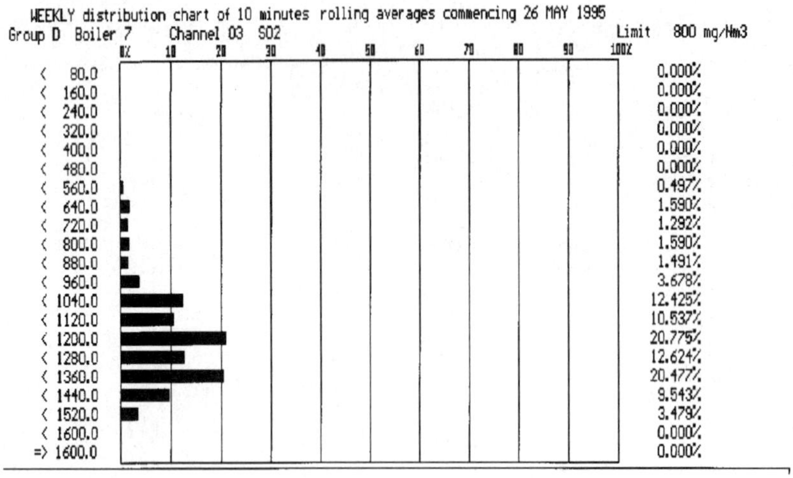

Fig. 14.12 PC display – data distribution report.

15

Quality assurance and quality control in emissions monitoring

Simon Medhurst and David Miles

15.1 INTRODUCTION

Emission monitoring is a relatively complex task. It will normally require a number of different items of equipment, and measurement of more than one parameter, and many of the techniques require collection of samples with subsequent transportation and laboratory analysis. An effective emission monitoring programme will produce reliable and accurate results, but there are many factors which can affect the validity of these results. Identifying these factors and effectively controlling the emission monitoring programme to produce reliable, valid results can be a complex task.

To ensure that we manage the emission monitoring programme effectively we apply techniques that produce good quality results. A number of terms such as quality assurance and quality control are used to describe the management procedures that we apply. Often these terms are confused. There are a number of standards employed in the UK relating to quality which is defined as 'the totality of features or characteristics of a product or service that bear on its ability to satisfy a given need'.

The British Standard definition of *quality assurance* is as follows:

> '. . . all those planned and systematic actions necessary to provide adequate confidence that a product or service will satisfy given requirements for quality. The definition of *quality control* is: the operational techniques and activities that are used to fulfil the requirements for quality.'

Many of the standards and quality systems and techniques have been initially developed from the manufacturing industry. It is much harder to measure the performance of a stack sampling programme than say, the production of bolts. In the latter case we can measure every bolt produced, reject every bolt out of tolerance, and our quality control department can tell us precisely the number of defects we are producing. With stack sampling there are many different processes involved – each of which should be assessed for quality – and it may not be obvious how this can be done. It becomes even more difficult if we extend our programme to include not only the delivery of the results but also the interpretation of the results and recommendations, i.e. the consultancy aspects other than simple monitoring.

The typical components of a quality assurance system are outlined in the BS 5750 checklist for quality systems (now BS EN 9001).

- Staff training
- Definition of management responsibility
- Contract review systems
- Document control
- Purchase control
- Traceability/audit trails
- Inspection and testing
- Review and inspection
- Control of non-compliances
- Formalized system of corrective actions
- Handling storage and delivery
- Quality records
- Quality audits

15.2 QUALITY MANAGEMENT FOR EMISSIONS MONITORING

The purpose of quality control is to reduce uncertainties in measurements to a minimum. Environment Agency Technical Guidance Note M2 (Environmental Agency, 1993) draws a distinction between the two types of uncertainty that arise in emissions monitoring. These may be classified as:

1. Random uncertainties – these give rise to a variation about the mean and tend to exhibit a Gaussian or similar statistical distribution. The degree of confidence that may be assigned to the mean of the measurements therefore increases as more samples are taken. Typically, random uncertainties arise from small variations in procedure and may be controlled by accurate specification of procedures and by training operators to ensure that these procedures are followed rigorously.
2. Systematic uncertainties – these introduce a bias into the measurements and do not vary about the mean. These can be reduced by ensuring the use of appropriate equipment, insisting on regular calibration and implementing a regular programme of maintenance. Systematic uncertainties also arise from differences in sampling

practice between operators. These latter uncertainties are the most difficult to trace and can only be corrected by training and vigilance in site audits.

Quantitative statistical aspects of handling these uncertainties are dealt with in Chapter 16. It is important to recognize that not all the uncertainty or variability of a measurement relates to 'errors' or 'mistakes' on behalf of the monitoring staff – for example, there will be the normal variability of the process and also limits to the accuracy of our measurement and analysis techniques which will contribute random 'noise'. Some of the most common sources of error and mistakes in emission monitoring will be reviewed in a later section. Some of the problems related to particle sampling have already been covered in Chapter 3 and issues relating to calibration and traceability are discussed in detail in Chapters 12 and 13.

While the various series of quality management standards (BS EN 9000, 9001, etc.), lay down widely applicable frameworks for setting up quality assurance systems, the most appropriate standard for testing laboratories is that specified by the National Accreditation for Measurement and Sampling (NAMAS) scheme. This is administered and regulated by United Kingdom Accreditation Service (UKAS) – the official UK national body in the field of laboratory accreditation.

The general criteria for calibration and testing laboratories are set out in NAMAS document M10 (NAMAS, 1989). This gives guidance on requirements for the establishment and maintenance of a quality system for laboratory testing (in this context a laboratory is defined as the party seeking NAMAS accreditation).

The definition of a quality system as set out in M10 is: the organizational structure, responsibilities, procedures, processes and resources for implementing quality management. The general requirements for this system are discussed in the following sections.

15.3 ORGANIZATION AND MANAGEMENT

Quality management requires a formalization of the job descriptions and responsibilities of the emissions monitoring team. This will include:

- Written job descriptions and definitions of training requirements and experience.
- Documented policies and procedures for the protection of proprietary rights and confidential information (for example, systems to ensure the protection of client confidentiality).
- The identification of an individual as a Technical Manager, with overall responsibility for the operation of the laboratory, and a Quality Manager responsible for the maintenance of the quality system on a day-to-day basis.

15.4 QUALITY SYSTEMS

The system of management for the maintenance of high standards and minimalization of error must be outlined in a Quality Manual. This should be maintained up to date and should contain:

- A statement of the policy of the organization with respect to the laboratory.
- A definition of the scope of the laboratory's testing activities.
- A description of the major equipment and techniques used.
- Organizational charts to define responsibilities.
- A list of approved signatories.
- A description of the procedures used to control all aspects of the laboratory's activities.
- A procedure describing the course of action to be followed on the discovery of deviations from those procedures. Where departures from procedure may be justified on technical or practical grounds, a record of the departure and its justification should be maintained.

15.5 STAFF

The Laboratory should only use staff for activities for which they have the appropriate training and qualifications and should only use staff employed by, or under contract to, the laboratory. In addition there should be:

- written definitions of the training requirements for each activity;
- policies and procedures to ensure that existing and new staff have and maintain relevant academic and professional qualifications; and
- records of training and experience for each staff member.

15.6 EQUIPMENT

An essential element of quality control is the use and maintenance of appropriate equipment. The NAMAS requirements specify:

- The use of equipment that is owned by, or on long-term lease or loan to, the laboratory (under normal circumstances).
- Each item of equipment should be uniquely identified.
- Regular maintenance and calibration of the equipment and the keeping of maintenance and calibration records.
- Careful segregation and labelling of malfunctioning or inaccurate equipment to avoid accidental use.

15.7 MEASUREMENT TRACEABILITY AND CALIBRATION

A critical element of the quality system is ensuring that the systems of calibration and measurement are 'traceable' to national standards of measurement and that confidence can be placed in the quality of measurements carried out at all steps in the traceability chain. This requires that the laboratory provides evidence that calibration or test procedures can be related to agreed standards. (For example, through the use of certified reference materials or by a suitable programme of measurement audits.)

All continuous emissions monitoring systems (CEMs) are susceptible to drift which may result in either an overestimate or an underestimate of the true concentration. These effects can be minimized, although not completely eradicated, by the strict application of a documented operation and maintenance procedure.

Calibration of equipment is a crucial element of this process. For CEMs it is necessary to draw a clear distinction between the routine calibration and a full-scale validation exercise. For routine purposes, a calibration carried out at zero and, preferably, two further points may be acceptable. Provided that this is carried out before and after testing, and the measured value lies within the bounds of the calibration intervals, then the result can be reported with confidence. Any persistent drift will also act as a guide to the changes which the instrument has undergone since its last full validation.

Validation is necessary to demonstrate the instrument's response over its full working range and should be carried out when the instrument is first installed, and at least annually thereafter. It should also be carried out whenever the instrument undergoes major repair maintenance. The purpose of validation is to relate the response of the instrument over its whole measurement range – usually a check of its linearity of response. This is carried out at, typically, 10% increments through each measurement range. Calibration checks should be taken in ascending and descending order throughout each range. Reference materials which are either certified standards themselves, or which can be related to certified standards – for example by the use of a gas divider – should be used.

Validation exercises are costly and time-consuming since sufficient data are necessary for statistical analysis which will provide a calibration function, regression line and confidence limits.

Since the calibration of an instrument determines the relationship between the instrument reading and the actual concentration of the pollutant of interest, it is an essential component of the quality control procedure and is, therefore, part of the routine operation of the instrument. It provides the link between the measurements which are being carried out and National Standards. From a quality assurance aspect the key element is the frequency of calibration and the use of appropriate certified reference materials (e.g. calibration gases).

15.8 SAMPLE STORAGE AND TRANSPORT

Quality assurance procedures for the storage and transport of samples are also addressed in detail in Technical Guidance Note M2 (Environment Agency, 1993). This account is drawn largely from that source.

The two main items to be addressed in respect of the storage and transport of samples are:

1. To ensure that the sample is identifiable throughout the sampling, sample preparation and analysis route.
2. To ensure that the sample integrity is maintained.

After collection, each sample should be given a reference number which identifies it throughout the system and all containers should be marked in such a way that the reference marks cannot be removed accidentally.

The reference number should be noted in a log which is specifically designed for the purpose. The log will provide details of:

- where the sample was taken;
- the date and time;
- the purpose of the sample;
- the method by which the sample was obtained; and
- the method of sample treatment.

If the sample is to be analysed by another organization, the date on which the sample was passed to the second organization should also be noted in the log. The suitability of the second organization to carry out the analysis should also be demonstrated (for example, by being NAMAS-accredited for the determinand in question).

Preservation of the sample integrity depends on the type of sample. Particulate samples such as those collected using the BCURA particulate sampler may be sealed after sampling and stored in suitable carrying cases. For most other samples, it is best to store them under cool conditions in the dark. Impingers can either be sealed, packed into cool boxes and transported to the laboratory for analysis or washed out into labelled containers and stored in cool boxes for transport back to the laboratory for analysis. Sample containers should be constructed of materials which are inert to the relevant determinand.

When samples are received at the laboratory, sample details should be entered into a second log book (in many cases the second log book will be a PC with relevant software). If the samples are not to be analysed immediately they should be stored in a cool dark place preferably in a refrigerator (imperative in the case of dioxin and PAH samples). The time for which a sample may be stored depends to a large degree on the type of sample and analysis required. For example, a sample of gas for chromatographic analysis may need to be analysed immediately, but a particulate sample for size analysis may be kept indefinitely if stored correctly. The length of time for which a sample can be kept depends on the reactivity of the analyte.

15.9 QUALITY AUDITS AND REVIEWS

Periodic auditing of the laboratory's activities is required to ensure that the stated policies and procedures are being followed. The planning and organization of the audits is the responsibility of the Quality Manager and the frequency of these should be as 'frequent as is necessary to achieve their purpose' and at least on an annual basis. An audit plan should be drawn up which schedules all audits in advance.

There are several types of audit and the procedure to be followed varies for each type of audit. The different types of audit may be summarized as:

- Horizontal audit
- Spot checks
- System audit
- Vertical audit

15.9.1 Horizontal audits

A horizontal audit covers one particular aspect of the quality system in detail, over the whole of laboratory.

Example – Horizontal audit of equipment

1. Check the calibration certificates for each item of equipment in that audit category. Check that the laboratory identification numbers agree with the labels on the equipment. Any equipment which is not in calibration must not be used for any accredited test.
2. Check that the calibration dates on the current certificate agree with the dates on the equipment labels.
3. Check that any data sheets and pro-forma used are current versions and that sufficient information is contained on them.
4. Check a selection of the items of equipment that have been calibrated and check the traceability of the measurements. Check the data sheets (if appropriate) and check the calculations. Ensure that the calculations have been checked and countersigned.
5. Check that the equipment used for calibration has been noted and check that these items of equipment were themselves in calibration at the time of the calibration.
6. Check that the technician carrying out the calibration has signed the certificate and that his/her training record is up to date.

15.9.2 Spot-check audits

These are, usually, not scheduled in advance but are carried out when it is believed that certain operational practices do not comply with either the quality control manual, in-house written procedures, standard test methods, or where appropriate, the requirements of NAMAS. These may occur, for example, after a customer complaint or following a routine audit.

In essence, a spot-check audit entails a check on the area of work involved, and the use of a spot-check audit check list to ensure coverage of all aspects of any problems that are identified.

When an audit has been carried out a report of the audit must be completed and any non-compliances dealt with by following the non-compliance procedure.

15.9.3 System audits

The purpose of a system audit is to examine work in progress to confirm that all aspects of the quality policy of the laboratory are being adhered to. The following guidelines should be used when carrying out a system audit.

1. Check that the staff carrying out the work have a training record which indicates that they are authorized to do so. Check that the operator is aware of his/her training record including limits of responsibility and authority.
2. Determine from the training records what degree of supervision should be present and verify that it is.
3. Check that the person performing the test is aware of the written procedures covering the work and that the current versions of these procedures are readily available.
4. Verify that the work is being done in accordance with the procedures.
5. Check that there are operating instructions available, if required, for the equipment in use and that the operator knows where to find them.
6. Check the validity of any reference materials (e.g. span gas) being used for calibration purposes.
7. Check the calibration traceability of items of equipment.
8. Check the test results for legibility and completeness.
9. Check any calculations on the current worksheet.
10. Assess the general housekeeping around the test area for anything which may prejudice the validity of the test in progress.
11. Check the availability of any protective clothing required.

15.9.4 Vertical audits

The purpose of a vertical quality audit is to trace the records of a single job backwards from the report. The records should demonstrate that the quality policy of the laboratory has been adhered to in all respects. The following guidelines should be used when carrying out a vertical quality audit.

1. **Report**: Examine the report for compliance with laboratory requirements (e.g. correct format).
2. **Calculations**: Check a selection of calculations for accuracy of mathematical calculations and correct transposition of data. When computer-calculated results are included, check that any manual data entry is accurate, and that the current software version is used.
3. **Original observations**: check that the correct data sheet has been used to record field measurements. Check that all original observations are legible, in ink, and complete. Check that any alterations have been done in a way that leaves the original figure readable and identifies the person making the change. Check field calculations.
4. **Equipment**: check that all measurement equipment is identified and that the calibration status has been checked before use. Trace the calibration documents for some of the items of equipment to verify traceability at time of test.
5. **Test results file**: check that the results file is complete and up to date. Make sure that all documents in the file are properly labelled so that they can be identified if they became misplaced.
6. **Staff**: check that all the paperwork has a signature indicating the person responsible for the work. Make sure that where required there is also a checker's signature. Refer to the staff training records to confirm that the staff who signed the records were properly authorized to carry out the work at *the time of signature.*

When an audit has been carried out an audit report sheet should be completed which records the details of the audit along with any non compliances. Details of the recommended corrective actions together with an agreed time scale for implementation and the name of the staff member responsible for making the corrective action should also be recorded on each sheet.

15.10 COMMON SOURCES OF ERROR

Table 15.1 draws together many of the most frequently encountered sources of error and the elements of the quality management system which should ensure that the problems are identified and corrected. In most cases the solutions lie in having correctly written protocols/procedures, adherence to guidelines laid down in a quality manual, correct selection and use of equipment and good staff training and experience.

Let us look in more detail at three of the individual areas which should be addressed in our programme.

Table 15.1 Quality management: sources of error

Error	Typical causes	Controlled by this area in quality system
Incorrect sampling point – may be process-related, i.e. sampling before all discharge points; or technique-related, i.e. sampling in turbulent conditions	Poor understanding of process	Quality control and quality assurance techniques – protocol training and application of correct measurements and procedures
Incorrect sampling technique	Cross-influences; over- or under-sampling, sample conditions exceed range of methods	Quality manual. Staff training/experience
Sample train leakages	Incorrect assembly tubing, etc., leaking under adverse weather or process conditions	Quality manual. Sampling protocol must always include leak checks
Inappropriate timing of sample run	Poor liaison between sample team and production	Quality manual. Staff training and experience
Inappropriate location of probe, time of sample or length of sample	Poor understanding of process and theory underlying sample protocol	Quality manual. Staff training and experience experience
Incorrect measurement of flow-rate and sample volume	Poorly maintained/calibrated equipment and use of unsuitable equipment	Quality manual. Equipment use and maintenance
Deposition of particulates in sampling probes, especially if external filtration is being used	Failure to wash sample lines adequately	Quality manual. Equipment use and maintenance
Adsorption on sample lines and in sample conditioning systems		Quality manual – use of heated sample lines of an appropriate material (e.g. Teflon)
Later desorption of compounds from sample lines, etc.	Poor sample storage and transport	Quality manual
Low collection efficiency in bubblers/impingers due to incomplete absorption	Selection of inappropriate absorption equipment	Quality manual
Low collection efficiency in absorbant tubes due to saturation (e.g. hydrocarbons)	Selection of inappropriate sample volumes	Staff training and experience – be aware of breakthrough volumes and do not draw too-large samples
Sample instability between collection and analysis	Use of inappropriate methods	Quality manual
Sample contamination	Dirty glassware, careless handling and transport, exposure to contaminants in the ambient atmosphere	Staff training

15.10.1 Sample position

The sample position – the ideal location – is described in Chapter 3. It is more important that airflow is uniform when sampling particulates. An ideal straight duct is hard to find, and often the location is compromised because of safe access requirements. Many processes in the UK which have not routinely been controlled by legislation have not been designed with sampling in mind, and provision for suitable site access has not been made available. Plant which have been regulated for a considerable period of time such as power stations normally have good access and sample positions. Often, many processes have more than one release point, and tracing all release points can be a complex exercise; for example, coating process ovens may release material at oven inlet and outlet points, as well as oven vents and drying areas. It sounds straightforward, but we must make sure that our sample location is downstream of all discharge sources.

Process guidance notes do not normally adequately describe the sample position. The sample position may also be influenced by the objectives of the sampling programme; for instance if we are investigating an odour problem and looking for potential release points our programme may include sampling at fugitive release points such as roof vents, doorways, etc., as opposed to just measuring the main extract duct which may be required for compliance testing. To ensure that we sample in the correct position, we need to be clear on the objectives of the programme and have a thorough understanding of the process.

15.10.2 Sample technique

Other chapters consider selection and utilization of correct sample technique and, although there have been some improvements, many sample techniques utilized for stack sampling may not have been fully validated under field conditions but have been developed from occupational uses when conditions are very different (lower concentrations and ambient temperatures and humidities).

Temperature and humidity can have a big effect on systems, reducing collection efficiency of absorption techniques utilizing charcoal tubes. It is important to research very carefully the background of the method to be used. Moreover, it may be necessary to undertake a trial to determine if the methods are acceptable.

15.10.3 Process understanding

Many sampling programmes achieve unrepresentative results as the sampling period chosen does not accurately represent the process emission. Many of the stack sampling techniques have been developed for relatively steady stack emissions such as power stations. It is not unusual in industrial processes to have a 100-fold difference in emissions over time periods from 10 months to 10 days. It is important to obtain as much information as possible relating to the process before commencing the supplier's programme. In practice, little data may be available concerning the process. Useful information can be obtained by examining raw material inputs, outputs, operating temperatures, etc., and initially to undertake a number of sampling runs to characterize the process adequately.

REFERENCES

Environment Agency (1993) Technical Guidance Note (Monitoring) M2, Monitoring emissions of pollutants at source (November, 1993), HMSO, London. ISBN 0-11-752922-2.

NAMAS Accreditation Standard M10 (1989). General Criteria Of Competence For Calibration And Testing Laboratories. United Kingdom Accreditation Service.

16

Statistics in relation to emissions monitoring

Andrew G. Clarke

16.1 INTRODUCTION

Specified emissions limits can provide a bewildering range of criteria:

'The emissions shall not exceed x mg/m^3'
'No more than 5% of the 48-hour averages should exceed y mg/m^3 in a calendar year'
'The calendar monthly average should not exceed z mg/m^3'
'95% of hourly averages for each rolling 24 hours should not exceed the limit value and the peak hourly value shall not exceed 1.5 times the limit value'
'15-minute mean emissions shall not exceed x mg/m^3 for more than 5% of the 15-minute periods of the process operation in any day and no 15-minute period shall exceed y mg/m^3'

Some of these criteria relate to continuous monitoring where the time resolution of the data may be as short as 1 minute or less but the period averages must then be computed (15-minute, hourly, daily, 48-hourly, monthly). Some relate to discontinuous monitoring which may be undertaken only a few times a year.

Pollution monitoring to meet these criteria can produce very considerable amounts of data which must be handled using statistical techniques. Familiarity with the techniques is important both for those producing the data and those in control authorities who must evaluate the significance of data submitted to them. The objective of this chapter is therefore to give a brief refresher course on the jargon and some of the techniques for those who have done statistics previously and, for those who are beginners, to provide a quick introduction and some signposts to further areas of study.

The areas which will be covered have had to be severely restricted. So, after some basic definitions and clarification of terminology, the primary targets are Confidence Limits and Tests of Hypotheses which involve either the Normal or the t -Distribution. This is standard material which is to be found in most statistics books or in the relevant parts of British Standard 2846 (Parts 1–3). The section on the statistics of percentile points of a distribution covers an area which does not seem to be well treated in the books but which, because of its relevance to so many of the emissions limits, is included here.

16.2 THE DATA

The variability of the data we collect may reflect

- the variability of the concentration in the flue, or
- the variability introduced by the measurement process (errors), or
- most likely, a combination of the two.

It will be important to bear this clearly in mind.

Consider the errors associated with the measurement process. Suppose you take 20 measurements and they are all within a few percent of the mean value. The *reproducibility* or *precision* is good. However, the *accuracy* may be poor because of

- losses in the sampling system;
- leaks in the sampling system;
- calibration errors, etc.

These effects give *systematic errors* which can be far larger than the *random errors* contributing to the observed variability. Systematic errors must be identified and eliminated by good experimental technique or by comparison of two systems based on different measurement principles (see Chapter 12). Statistics usually concerns only the random variability.

Now consider the actual time variations in flue gas concentration (Fig. 16.1). On a short time scale there are random variations from the mean value, but superimposed on this is a longer-term drift in concentration – initially up and later down. The drift may relate to minor changes in process conditions but, however it arises, it is unlikely to be random in nature. One therefore has to be careful in the application of statistical techniques where the assumption that the variations are random is often made.

16.3 FUNDAMENTALS

If we have a set of N measurements – the **sample** of data – then the arithmetic mean is

$$\overline{x} = \sum_{\mathrm{i}} x_{\mathrm{i}}/N \tag{16.1}$$

where the index i runs from 1 to N and indicates the individual measurement.

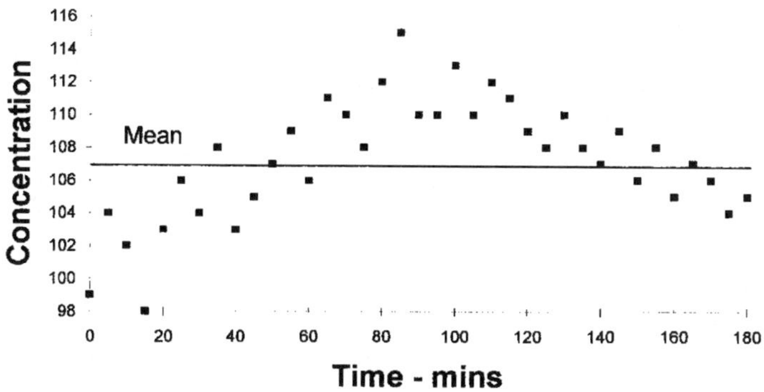

Fig. 16.1 Illustration of random and apparently systematic variations in a time series of data.

The **variance** is

$$s^2 = \sum_i (x - x_i)^2 / (N-1) \qquad (16.2)$$

and the **standard deviation** *s* is the square root of the variance.

Note that the divisor in Eqn (16.2) is taken to be $(N - 1)$ rather than N. Calculators often give both options for the calculation. The reason for this concerns the relation between the **sample** (the N values actually measured) and the hypothetical **population** (infinite number of potential measurements) from which the sample is considered to be drawn. While the arithmetic mean $\bar{x}$ (Eqn (16.1)) is the best, unbiased estimator of the population mean μ, the sample standard deviation s calculated using N gives a biased estimate of the population standard deviation (denoted σ). Equation (16.2), using $(N - 1)$ gives an unbiased estimate and should normally be used. Obviously, the difference is only significant for small sample numbers.

The standard deviation is a measure of the scatter or dispersion of the data about the mean. As we shall see later, a rough guideline for random data that is normally distributed is that 95% of the points lie within 2 standard deviations of the mean. Another way of describing the dispersion is to express the standard deviation as a percentage of the mean value which is called the **Coefficient of Variation** usually expressed as a percentage:

$$V = 100(s/\bar{x}) \qquad (16.3)$$

The accuracy with which the mean value is determined actually increases with increasing sample number. The appropriate measure of uncertainty of the mean is the **standard error** defined by

$$s_{\bar{x}} = s/\sqrt{N} \qquad (16.4)$$

This is an estimate of the standard deviation of the population of the mean values calculated for samples of size N. The standard error decreases by a factor of 2 for an increase in number of data by a factor of 4.

Example 1: Means, Standard Deviations and Standard Errors

Suppose we have measured 1-minute averages of the emissions of a continuous process operating at steady conditions for 1 hour. The total set of data (60 points) lies within the range 37 to 65 mg/m^3 and has a mean of 50 and a standard deviation of 6. The standard error of that mean would be $s_{\text{hour}} = 6/\sqrt{60} = 0.775$. If we monitored for the whole day (1440 points) we might find the range covered from 35 to 70 mg/m^3 but had the same mean value of 50 and the same standard deviation. The standard deviation expresses the spread or dispersion of the values about the mean and does not vary significantly with the number of data. On the other hand, the standard error would now be $s_{\text{day}} = 6/\sqrt{1440} = 0.158$. The more values that are included in the mean, the more closely defined is its value. This assumes that the full range of variability might be experienced within any hour of the day. In practice, this might not be the case.

The **frequency distribution** of the data is often assumed to be **normal** providing the standard deviation is only a small fraction of the mean value. Normality implies a symmetrical distribution of data points about the mean and many of the statistical tests assume approximate normality. However, there is no particular reason why normality should apply. Analysis of various experimental data sets for dust emissions by Harding *et al.* (1994) showed that they were far from ideal and that there were often short-term fluctuations to high values far from the mean. In other words, the distribution is skewed to higher values.

Normality can be tested by plotting the cumulative frequency distribution on probability paper. Significant curvature indicates deviation from normality. Similar paper is available for **log-normal** data, i.e. where the logs of the concentration values follow a normal distribution. Figure 16.2 illustrates a hypothetical set of data that is log-normally distributed both as a histogram of the frequencies of occurrence of particular concentration values and also as a cumulative frequency plot. Note that with the log-normal distribution the most frequently occurring value is less than the geometric mean value (obtained by averaging the log(concentration) values) which in turn is less than the arithmetic mean. The arithmetic mean is well to the right of the peak in the distribution.

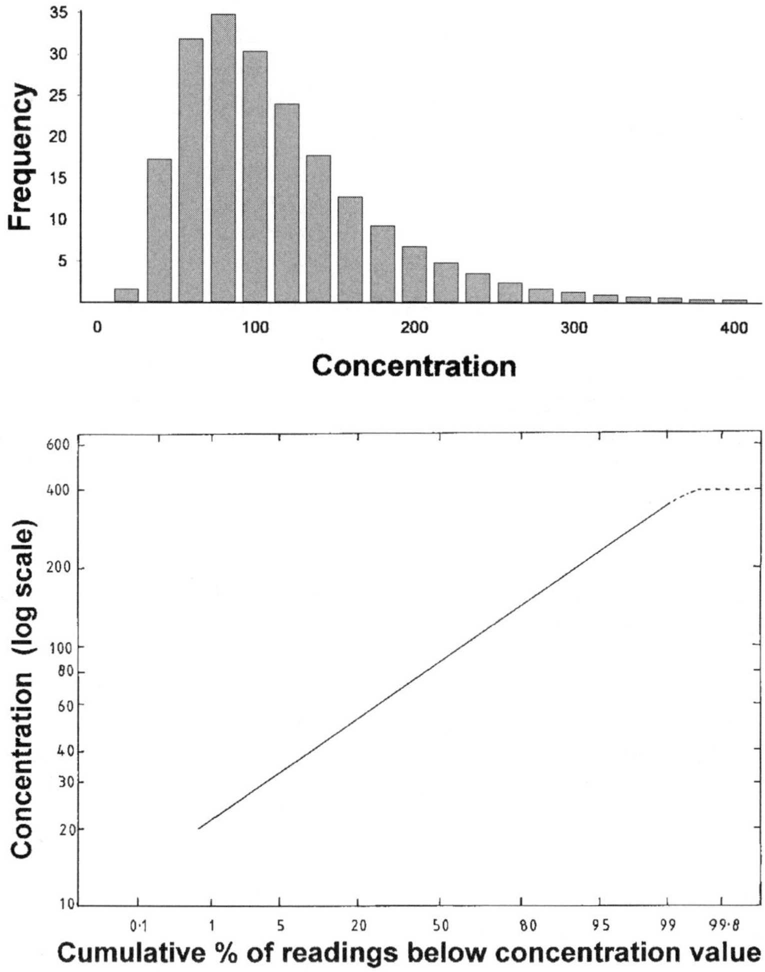

Fig. 16.2 (a) Distribution of data which is distributed log-normally. There are 200 points with geometric mean 100, arithmetic mean 113, and geometric standard deviation 1.7. (b) Cumulative frequency distribution of the data in (a).

The emissions criteria often specify the 95% point for continuous monitoring situations. However, this point may depend on only a very small number of readings. For example, there are only 168 hours in a week, so 5% of hourly readings corresponds to 8.4 readings. Eight exceedances of the limit in a week and you pass, nine exceedances and you fail. The situation is worse for 15-minute averages over a day (96 readings). 5% corresponds to only 4.8 readings – five exceedances and you fail. In these situations it is valuable to plot the cumulative frequency distribution as in

Fig. 16.2 because it may indicate whether the top 5% of readings are consistent with the distribution in the 50–95% range or whether they look like an untypical excursion to higher or lower values. Quantitative treatment of the uncertainty associated with the percentile points of a distribution is dealt with in section 16.7 below.

16.4 CONFIDENCE LIMITS

16.4.1 Using the normal distribution

For any particular statistic that we calculate it is useful to be able to estimate the degree of uncertainty or error bounds. If we assume that the population is approximately normally distributed then we can predict the frequencies of occurrence of particular values relative to the mean. Figure 16.3 shows the normal distribution. The horizontal axis is defined in terms of the number of standard deviations from the mean usually denoted z:

$$z = (x - \mu)/\sigma \tag{16.5}$$

where μ is the population mean and σ the population standard deviation. The upper and lower 2.5% of the area are shaded so that 95% of the population values occur in the defined band. The lines are at $z = \pm 1.96$. We thus affirm that, with 95% confidence level, individual values should fall within 1.96σ of the mean. Since we do not know the population mean and standard deviation, we estimate them using the sample mean and standard deviation (Eqns (16.1) and (16.3)). The corresponding values of z for different confidence levels are shown in Table 16.1.

Example 2: Confidence limits on means

If 36 measurements result in a mean of 80 and standard deviation 18, then the standard error of the mean is $18/\sqrt{36} = 3$. We use the standard error because we are considering the population of possible mean values. With 95% confidence the true mean lies in the range $80 \pm 1.96 \times 3$ i.e. 74.1 to 85.9. For 90% confidence, the range is 1.645×3 or 75.1 to 84.9.

16.4.2 Using the t-distribution

The normal distribution is appropriate if the sample number is relatively large, say $N > 30$. For small samples it is not appropriate and it is necessary to use the t-distribution. Like the normal distribution, to which it is identical for large sample numbers, the t-distribution is symmetrical

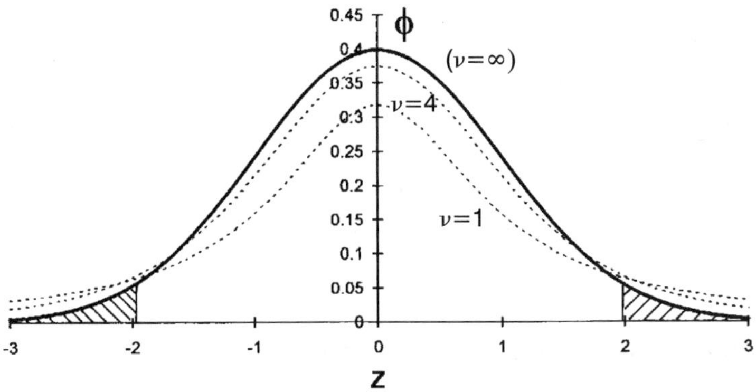

Fig. 16.3 The normal distribution (solid curve). The upper and lower 2.5% of the area are shaded so that 95% of the population values occur in the defined band. The dashed curves show the t-distribution for comparison using 1 and 4 degrees of freedom.

Table 16.1 Comparison of the normal and t-distribution

		Percentile (2-tailed)				
		99	98	95	90	80
Normal Distribution	z	2.58	2.33	1.96	1.645	1.28
t-distribution						
Degrees of freedom (ν)	t					
∞		2.58	2.33	1.96	1.645	1.28
30		2.75	2.46	2.04	1.70	1.31
20		2.84	2.53	2.09	1.72	1.32
10		3.17	2.76	2.23	1.81	1.37
5		4.03	3.36	2.57	2.02	1.48
2		9.92	6.96	4.30	2.92	1.89
1		63.66	31.82	12.71	6.31	3.08
Percentile (1-tailed)		99.5	99	97.5	95	90

about the mean value; however, it is flatter and broader (Fig. 16.3 and Table 16.1). This means that for any particular confidence level the confidence limits are rather broader than with the normal distribution. What is more, they vary according to how large the sample is or, more precisely, with the number of **degrees of freedom** ν. When considering means the number of degrees of freedom is simply $\nu = N - 1$. The confidence limits on the true mean are now, for example

$$\bar{x} - \frac{t_{95}s}{\sqrt{N}} < \mu < \bar{x} + \frac{t_{95}s}{\sqrt{N}}$$

where t_{95} is the t value for 95% confidence from Table 16.1 using the appropriate number of degrees of freedom.

Example 3: Confidence limits – small sample numbers

Let us take an extreme example of small sample numbers. BS 3405 requires duplicate measurements of particulate emissions with the condition that the higher reading should not be more than 50% above the lower reading. Suppose we make two measurements of 60 and 90, mean 75 mg/m^3. These values are at the limit of acceptability.

Sample number N	= 2
Degrees of freedom ν	= 1
Standard deviation s	$= [((15)^2 + (15)^2)/(2{-}1)]^{0.5}$
	= 21.2
Standard error $s/\sqrt{N}$	$= 21.2/\sqrt{2} = 15$

Referring to Table 16.1, the 80% confidence limits on the mean for 1 degree of freedom are calculated using a t value of 3.08. The confidence limits are thus 3.08 times the standard error, i.e. ± 46.2. That is, there is a 20% chance that the true value lies outside the range 28.8 to 121.2. There may be good arguments why we may believe that we have actually done rather better than this in defining the emissions but that is the appropriate statistical estimate! With only two samples we know virtually nothing of the distribution of values and this example should be taken as illustrating the approach rather than giving useful statistical information!

16.5 HYPOTHESIS TESTING

The probabilities associated with the normal and t-distributions can be used in a slightly different way to test specific hypotheses concerning the data. From the sample parameters and the level to be tested we evaluate the z or t value and compare it with the critical value for particular degrees of confidence. We can then accept or reject the hypothesis with those levels of confidence.

Example 4: Exceedance of an emissions limit

Taking the same data as in Example 3, namely two readings of 60 and 90 mg/m^3, let us suppose that the emissions limit is $L = 100$ mg/m^3 and test the hypothesis that the true (population) mean μ is less than 100. We calculate the t value:

$$\begin{aligned} t &= (L - \bar{x}) / s \qquad (16.6) \\ &= (100 - 75)/15 \\ &= 1.67 \end{aligned}$$

The critical values of t for 1 degree of freedom from Table 16.1 are

	t_{cr}
90% confidence	3.08
80%	1.38
75%	1.00, etc.

In evaluating probabilities associated with an upper limit we have what is called a '**one-tailed**' test rather than the '**two-tailed**' test used for confidence limits. The value of 3.08 is associated with 90% confidence here rather than 80%, as in Example 3.

Since $|t| < 3.08$, we reject the hypothesis that the mean is less than 100 at 90% confidence. However, we would have to accept it at the 80% level since 1.67 > 1.38. For these data points with a mean of 75 there is a chance of between 10 and 20% that the true mean is greater than the limit value. Again, it should be emphasized that one would hardly use a test of this sort with only two data points and the example is used simply to illustrate the method.

16.6 COMPARISON OF MEANS

There are a number of situations where we may wish to test whether two mean values are significantly different or not. One example might be the intercomparison of two different techniques for monitoring the emissions of the same species. Another might be two series of tests before and after some modifications have been made to a plant. Has there been significant improvement or not?

Example 5: Intercomparison of methods

Using data drawn from the author's studies on atmospheric pollution (Clarke *et al.*, 1977), three methods of measuring particulate concentrations were compared on the basis of daily averages. The results were as follows:

	Sample number N =	28
Overall averages	Method A	74 μg/m³
	Method B	70 μg/m³
	Method C	86 μg/m³

In this case we have the same number of data points for each method and they were take over the same time periods (side-by-side comparison). Since there is a large day-to-day variability of atmospheric conditions the individual standard deviations are quite large and a comparison of means based on these is not very helpful. We use the method of paired comparisons based on the daily differences between the mass concentrations M taken two at a time:

$$d_j = M_{Aj} - M_{Bj} \tag{16.7}$$

The mean value of these differences is $\bar{d}$ and the variance

$$s_d^2 = \sum_j (d_j - \bar{d})^2 / (N-1) \tag{16.8}$$

The t- value is then

$$t = \bar{d}\sqrt{N} / s_d \tag{16.9}$$

The hypothesis is tested that the average difference between methods $\bar{d}$ is significantly different to zero. The number of degrees of freedom is $(N - 1) = 27$ and the results were calculated to be

Methods A and B $t = -1.14$
Methods A and C $t = 2.91$

The critical value of t for 95% confidence is 2.05 (two-tailed test, Table 16.1). The difference between methods A and B is therefore not significantly different to zero at 95% confidence whereas the difference between methods A and C appears to be significantly different at this level of confidence.

It would have made very little difference in this case ($N = 28$) if we had used the normal distribution to carry out the test since the critical value would then have been 1.96 instead of 2.05.

In the case of two groups of measurements taken at different times on a particular plant the data are not paired and there may not be the same number of values in the first series of tests as in the second:

Before: Emissions rate $\overline{E}_1$ calculated from N_1 values with standard deviation s_1

After: Emissions rate $\overline{E}_2$ calculated from N_2 values with standard deviation s_2

We make a pooled estimate of the variances:

$$S_{12} = \left[\frac{(N_1 - 1)s^{12} + (N_2 - 1)s_2^2}{(N_1 + N_2 - 2)}\right]^{1/2} \qquad (16.10)$$

and calculate the test statistic

$$t = \frac{\overline{E}_1 - \overline{E}_2}{s_{12}[N_1^{-1} + N_2^{-1}]^{1/2}} \qquad (16.11)$$

The number of degrees of freedom are $(N_1 + N_2 - 2)$. We again test the hypothesis that the difference in the means is not significantly different to zero (which would be a two-tailed test allowing either positive or negative differences). Alternatively, we could test the hypothesis that the 'after' readings are significantly *lower* than the 'before' readings, which would be a one-tailed test.

Example 6: 'before' and 'after' comparison

Suppose:	Before:	$\overline{E}^1 = 65$	$N_1 = 5$	$s_1 = 3.5$
	After:	$\overline{E}^2 = 61$	$N_2 = 8$	$s_2 = 2.7$

The combined standard deviation is $s_{12} = 3.02$
The calculated t value is $t = 2.32$
Degrees of freedom = 11

For 95% confidence in a one-tailed test and 11 degrees of freedom, $t_{cr} = 1.8$. Since the actual t is greater than t_{cr}, we can say that the difference is significantly greater than zero (i.e. the 'after' readings are significantly lower than the 'before' readings) at the 95% confidence level.

16.7 STATISTICS RELATING TO THE PERCENTILE POINTS OF A DISTRIBUTION

As was noted earlier, the 95% point of a distribution may, in practice be determined from a very small number of data – often in single figures. It is therefore subject to considerable uncertainty. A set of data which are approaching an emissions limit expressed in terms of the 95% point will stand a significant chance of being classified as satisfactory when further data would prove that it was unsatisfactory, and vice versa. The problem was recognized by the National Rivers Authority in their determination of water quality classes for rivers (National Rivers Authority, 1991). The class definitions depend on such factors as the 95% level of the Biological Oxygen Demand (BOD). For example, the 1b class requires 95% of the BOD values to be below 5 mg/l. A river which has a true 95% value of 4.2 mg/l is clearly in this class; however, if there were only 36 analyses available they estimated that there would be a 17% chance that it would be wrongly classified as being worse than it actually is (i.e. >5 mg/l). In the more recent General Quality Assessment Scheme 90 percentile values are used instead of 95 percentiles which reduces the problem significantly.

This problem can be partly handled using probability theory. Consider a population of data such that 95% of the values are below a critical value (the emissions limit) and, correspondingly, less than 5% exceed the critical value. If we consider sampling randomly from that population, then the probability of an individual measurement exceeding the limit is 5% ($p = 0.05$) and of falling below the limit 95% ($1 - p = 0.95$). For a given set of N measurements the distribution of the measured number of exceedances X will be determined by the binomial distribution and the results are shown in Table 16.2.

With only 20 measurements there is a 64% chance that we would measure one or more exceedances (and claim that the limit was breached) and a 36% chance that no exceedances would be detected. With 100 measurements, there is a 56% chance that five exceedances or more would be detected. In the limit of large sample numbers there would be an equal chance of the measured number of exceedances being above or below the ideal number corresponding exactly to the 5% probability in the defined population.

What is perhaps instructive is the very wide spread of the number of exceedances which might arise in this situation. So, with 200 points and the ideal number of exceedances being 10 (5%) for the case we have taken, there is in fact at least a 1% chance of us actually getting any number between 3 and 17 corresponding to estimated probabilities of 1.5 and 8.5%. In fact, for large sample numbers the distribution illustrated in Table 16.2 becomes approximately normal with standard deviation

$$s_p = [p(1-p)/N]^{1/2} \qquad (16.12)$$

Table 16.2 Probabilities (%) of measuring a given number *X* of exceedances of a limit value in a total of *N* measurements when the theoretical probability of exceedance is 5%

X	*N*			
	20	50	100	200
0	36	8	1	
1	38	20	3	
2	19	26	8	
3	6	22	14	1
4	1	14	18	2
5		7	18	4
6		3	15	6
7			11	9
8			6	11
9			3	13
10			2	13
11			1	12
12				10
13				7
14				5
15				3
16				2
17				1

This is the square root of the variance of sample proportion which is covered in most statistics books. The analysis of course focuses on the two simple categories Pass/Fail and it gives us no information about the distribution of data about the emissions limit value. In fact, one of the advantages of using limits based on the percentile points is that no assumptions need to be made about the form of the distribution.

Example 8: Uncertainty of the percentile point (or sample proportion)
Suppose we have 200 readings of a gas concentration for which the emissions limit is specified that 95% of the readings are to be below 75 μg/m³. The highest 20 readings are as follows:

69	4 readings	75	2
70	2	76	1
71	1	77	1
72	3	78	2
73	2	81	1
74	1		

We note that only seven points (3.5%) are above 75 and that the 95th percentile point is 72 μg/m³. The data satisfy the set criterion

emissions. But how confident are we that a larger number of points would not lead to the conclusion that over 5% of the data exceed the limit value? In other words, we test the hypothesis that the true sample proportion is less than 0.05.
Using the formula above, the probability $p = 0.035$ and the standard deviation of the probability (or sample proportion) is

$$s_p = [0.035 \times 0.965/200]^{1/2}$$
$$s_p = 0.013$$

We calculate the value

$$z = (\text{test value of proportion} - \text{observed value})/\, s_p = 1.2.$$

(A change of p from 0.035 to 0.05 represents an increase of 1.2 standard deviations.) From Table 16.1, the likelihood of this not occurring is 90% (one-tailed test with $z = 1.28$). So, there is only a 10% chance that more than 5% of the points would exceed the limit if we took a lot more data. An alternative use of the same information is to say that the confidence limits on our fraction exceeding the limit is

$0.035 \pm 1.28 \times 0.013$ or 0.018 to 0.052 (1.8 to 5.2%) with 80% confidence.

Note: this is a two-tailed test.
Given the small number of exceedances, the normal approximation is not very good in this case. Confidence limits for the estimated sample proportion for small sample numbers are given by Davies and Goldsmith (1972, pp. 445–7) and would lead to more accurate estimates, particularly for the lower limit, and the confidence band would not be symmetrical above and below the observed proportion.

We now consider the uncertainty of the concentration values associated with particular points on the cumulative frequency distribution. For very large sample numbers the sampling distributions are approximately normal and we can use the same approach to confidence limits and tests of hypothesis previously described. The appropriate standard errors are (Spiegel, 1972, p. 144; Patel and Reed, 1982):

Median	50%	$1.2533\sigma\sqrt{N}$
	75%	$1.3626\sigma\sqrt{N}$
	80%	$1.4288\sigma\sqrt{N}$
	90%	$1.7094\sigma\sqrt{N}$
	95%	$2.1137\sigma\sqrt{N}$

Note: the numerical values in this table are $[p\,(1-p)]^{1/2}/\phi$ where $p = 0.1$ if we define p as the fraction above the 90% point for example and ϕ is the ordinate of the normal curve of error (Fig. 16.3) for the z value corresponding to p; e.g. $z = 1.28$ for 90%, $\phi = 0.176$.

Note that the standard error of the median is 25% larger than the standard error of the mean and that as you move out towards the tail of the distribution the standard error progressively increases. As a rough guide if we have large sample numbers, the confidence limits on the 95% point will be twice as wide as those on the mean. For smaller sample numbers the limits may be wider still.

Example 9: exceedance of a 95% limit

Suppose we have the same 200 data points as in Example 8, but we consider the additional information that the mean value is 57 and the standard deviation 8 $\mu g/m^3$. The 95% value is found to be 72 (this corresponds to the 11th largest value out of the 200) which is just below the emissions limit of 75. This data set satisfies the legal requirement but test the hypothesis that the true 95% point is less than the limit value.

The estimated standard error of the mean $s/\sqrt{N} = 0.57$, so the standard error of the 95% point will be $2.1137 \times 0.57 = 1.20$. The calculated z value is

$$z = (75 - 72)/1.20 = 2.49$$

The critical z value for 99% confidence in a one-tailed test (Table 16.1) is 2.33. Since this is less than the experimental value we confirm the hypothesis at this high level of confidence. The data satisfy the emissions criterion without significant statistical uncertainty.

Comparing Examples 8 and 9, it will be seen that in the first case the only information considered is the total number of points and the number exceeding the limit. The statistical uncertainty associated with the fraction of exceedances is examined assuming random sampling out of a constant but unknown distribution. In Example 9, it is possible to assess the uncertainty of the position of the 95% point relative to the limit because we have supplied some additional information about the distribution, namely its mean and standard deviation. That test does seem to give a greater degree of confidence in the data but there is the underlying assumption about normality which, as has been emphasized previously, is probably not correct.

REFERENCES

British Standard 2846 – Guide to the statistical interpretation of data.

- Part 1, 1991, Routine analysis of quantitative data
- Part 2, 1985, Estimation of the mean: confidence interval (= ISO 2602)
- Part 3, 1985, Determination of a statistical tolerance interval (= ISO 3207)
- Part 4, 1985, Techniques of estimation and tests relating to means and variances (= ISO 2854)
- Part 5, 1985, Powers of tests relating to means and variances (= ISO 3494)
- Part 6, 1985, Comparison of two means in the case of paired data (= ISO 3301)
- Part 7, 1993, Tests for departure from normality (= ISO/DIS 5479)
- Part 8, 1995, Estimation of a median

Clarke, A.G., Moghadassi, M.A. and Williams, A. (1977) *Journal of Aerosol Science*, **8**, 73.

Davies, O.L. and Goldsmith, P.L. (1972) *Statistical Methods in Research and Production*, 4th edn, ICI Publication, Oliver & Boyd, Edinburgh.

Harding, E.F., Griffiths, R.F. and Sharratt, P.N. (1994) *Statistical Aspects of Monitoring*. Report produced for DoE/HMIP, RR/94/006, January.

National Rivers Authority (1991) *The Quality of Rivers, Canals and Estuaries in England and Wales*, Report of the 1990 Survey, p. 53.

Patel, J.K. and Read, C.B. (1982) *Handbook of the Normal Distribution*, Marcel Dekker, New York, p. 262.

Spiegel, M.R. (1972) *Statistics*, Schaum's Outline Series, McGraw-Hill.

Appendix A

British Standards

PARTICLES

Extractive sampling:

BS 3405, 1983. Measurement of particulate emission including grit and dust (simplified method)
BS 893, 1978. Measurement of the concentration of particulate matter in ducts carrying gases.
BS 6069, The characterization of air quality;
Part 4, Stationary source emissions
Section 4.2, 1991. Method for the determination of asbestos plant emissions by fibre count measurement.
Section 4.3, 1992. Method for the gravimetric determination of concentration and mass flow rate of particulate material in gas-carrying ducts.

Smoke:

BS 2811, 1969. Specification for smoke density indicators and recorders.
BS 2742, 1969. Notes on the use of the Ringelmann and miniature smoke charts.
BS 2742, Addendum No 1., 1972. The calibration of instruments in Ringelmann Number.

Particle size:

BS 3406 Methods for the determination of particle size distribution.
Various parts describing different methods including microscopy (Pt.4, 1993)and light scattering (Pt.7, 1988) but none specifically for flue gases.
BS 6069 The characterization of air quality, Section 3.5, 1993 Workplace atmospheres – size fraction definitions for measurement of airborne particles (also known as BS EN 481, 1993)

GASES

Manual methods:

BS 1756 Methods for Sampling and Analysis of Flue Gases,
Part 1, 1971. Methods of sampling

Part 2, 1971. Analysis by the Orsat apparatus.
Part 3, 1971 Analysis by the Haldane apparatus.
Part 4, Miscellaneous Analyses (1977).

1. Scope
2. References
3. Determination of moisture
4. Determination of oxides of sulphur: propan-2-ol method
5. Determination of oxides of sulphur: direct method
6. Determination of sulphuric acid dewpoint

7. and 8. Determination of carbon monoxide: iodine pentoxide method

9. Determination of oxides of nitrogen: modified Saltzman method.
10. Determination of oxides of nitrogen: phenol disulphonic acid method.

BS 6069, The characterization of air quality;
Part 3, Workplace Atmospheres.

Section 3.1, 1989. Method for the determination of vinyl chloride using a charcoal tube and a gas chromatograph.

Section 3.2, 1991. Method for the determination of particulate lead and lead compounds by flame atomic absorption spectrometry.

Section 3.3, 1991. Method for the determination of chlorinated hydrocarbons by charcoal tube/solvent desorption/gas chromatography.

Section 3.4, 1991. Method for the determination of various aromatic hydrocarbons by charcoal tube/solvent desorption/gas chromatography.

Section 3.5 – see under Particles above.

Section 3.6, 1994. General requirements for the performance of procedures for the measurement of chemical agents (also known as BS EN 482, 1994).

Part 4 Stationary source emissions

Section 4.1 Method for the determination of the mass concentration of sulphur dioxide: hydrogen peroxide/barium perchlorate/Thorin method. (=ISO 7934, 1989)

Sections 4.2 and 4.3 – see under Particles above.

Section 4.4, 1993. Determination of the mass concentration of sulphur dioxide – performance characteristics of automated measuring methods.(=ISO 7935, 1992)

Part 5 Characterization of air quality. General.

Section 5.1, 1994 Handling of temperature, pressure and humidity data. (=ISO 8756, 1994)

Continuous methods:

BS 3048, 1958. Code for continuous sampling and automatic analysis of flue gases: indicators and recorders.
BS 4314 Specifications for apparatus for physical methods of gas analysis.
Part 1, 1968. Infra-red analysers for industrial use.

Calibration gases:

BS 4559 Methods for preparation of calibration gas mixtures.

Appendix B
German VDI Guidelines (Verein Deutscher Ingenieure)

These can be ordered from: Beuth Verlag, Postfach 1145, Burggrafenstrasse 6, D - 1000, Berlin 30, Germany. Tel: (49) 30 2601 2260, Fax: (49) 30 2601 1231. For recently published methods English versions are available as indicated.

Compound	**Guideline/ Sheet Number**	**Method**	**Year**	**English**
Planning of spot sampling measurements of stationary source emissions				
	2448/1		92	✔
Calibration of Automatic Emissions Monitoring Systems				
	3950/1		94	✔
Calibration Gases				
	3490 sheets 1–17 cover methods of preparation of calibration gas mixtures.			
Gases				
$NO + NO_2$	2456/1	phenol disulfonic acid reaction	73	
$NO + NO_2$	2456/2	titration after oxidation with H_2O_2	73	
NO_2	2456/4	UV absorption (Limas G)	76	
NO	2456/5	Chemiluminescence (Thermo Electron Model 10)	78	
$NO + NO_2$	2456/6	analysis of NO with converter, NDIR or chemiluminescence analyser	78	
NO	2456/7	chemiluminescence (atmospheric pressure instrument)	81	
$NO + NO_2$	2456/8	sodium salicylate method	86	✔
NO	2456/9	NDUV resonance (RADAS 1)	89	✔
$NO+NO_2$	2456/10	dimethylphenol method	90	✔
Organic compounds	2457/1	GC principles	91	
1,1,1 trichloro-ethane	2457/2	GC	74	
Trichloro-ethylene	2457/3	GC	76	
Tetrachloro-ethylene	2457/4	GC	75	
Benzene, toluene, xylene	2457/5	GC	81	
Butyl acetate (*n-*, *iso-tert,-*)	2457/6	GC	81	

Acetic acid esters	2457/7	GC	81	
CO	2459/6	Non-Dispersive IR	80	
CO	2459/7	iodine pentoxide method	94	✔
Organic compounds	2460/1	IR, principles	92	
Dimethyl-formamide	2460/2	IR	74	
Cresols	2460/3	IR	81	
SO_2	2462/1	iodine/ thiosulphate method	74	
SO_2	2462/2	H_2O_2 method, titration with NaOH	74	
SO_2	2462/3	H_2O_2 method, grav. analysis as $BaSO_4$	74	
SO_2	2462/4	IR UNOR 6 + URAS 2	75	
SO_2	2462/5	conductivity method (Mikrogas-MSK-SO_2 E1)	79	
SO_2 calibration	2462/6	Calibration of automatic SO_2 systems on combustion plant	74	
SO_2	2462/7	2-propanol method	85	✔
SO_2	2462/8	H_2O_2 method, titration with $BaCl_2$ against Thorin	85	✔
Hydrocarbons	2466/1	IR absorption (URAS 1 and 2)	73	
Fluorine compounds	2470/1	absorption in caustic soda	75	
HCl	3480/1	absorption (low Cl particulates)	84	✔
HCl	3480/2	continuous detection. – SPECTRAN 677IR	92	✔
Gaseous inorganic chlorine compounds	3480/3	continuous detn. - ECOMETER	92	✔
Hydrocarbons	3481/1	Flame Ionization Detector	75	
Hydrocarbons	3481/2	absorption on silica gel	80	
Organic solvents	3481/3	Flame Ionization Detector	92	
Total organic carbon	3481/6	Choice and application of methods	94	✔
H_2S	3486/1	potentiometric titration	79	
H_2S	3486/2	iodometric titration	79	
H_2S	3486/3	colorimetric method	80	
CS_2	3487/1	iodometric titration	78	
Cl_2	3488/1	methyl orange method	79	
Cl_2	3488/2	bromide–iodide method	80	
Vinyl chloride	3493/1	GC	82	
Basic nitrogen compounds	3496/1	absorption in sulphuric acid	82	
Polychlorinated dibenzo dioxins and furans	3499/1E	dilution method	90	

Polychlorinated dibenzo dioxins and furans	3499/2E	filter/cooler method	90	
Aliphatic aldehydes C1–C3	3862/1	MBTH method	90	✔
Acrylonitrile	3863/1	gas chromatography, grab sampling	87	✔
Acrylonitrile	3863/2	gas chromatography, absorption in low-temperature solvents	91	✔
Acrylonitrile	3863/3E	absorption on activated carbon, desorption by dimethyl formamide	88	
Polycyclic aromatic hydrocarbons (PAH)	3872/1	engine emissions by GC	89	✔
hydrocarbons (PAH)	3782/2E	diluted Otto + diesel exhausts GC	89	
hydrocarbons (PAH)	3783/1	stationary industrial plant – dilution method (RWTÜV) with GC	92	✔
1,3 butadiene	3953/1E	absorption onto activated carbon, GC	91	

Calibration of Automatic Emissions Monitoring Systems

	3950/1E		91	

Particles

	2066/1	gravimetric detection of dust loading. Survey	75	
	2066/2	manual dust measurement, tubular filter devices 4–12 m^3/h	93	✔
	2066/3	manual dust measurement, tubular filter devices, 40 m^3/h	94	✔
	2066/4	continuous optical transmissometry	89	✔
	2066/5	fractionated dust measurement cascade impactor	94	✔
	2066/6	continuous, scattered light photometer KTN	89	✔
	2066/7	gravimetric – low dust content plane filter method	93	✔
	2268/1	chemical analysis of particulate emissions Ba,Be,Cd,Co,Cr,Cu,Ni,Pb,Sr,V,Zn by atomic spectrometry	87	✔
	2268/2	As,Sb,Se atomic abs. spectrometry after separation of volatile hydrides	90	✔
	2268/3	Tl by atomic abs. spectrometry	88	✔
	2268/4	As,Sb,Se by graphite furnace A.A.	90	✔
	3868/1	total emissions of metals, metalloids and their compounds, manual method	92	✔

Appendix C

US EPA Standard Methods

Details in the Code of Federal Regulations – Protection of the Environment, 40, Part 60 Appendix A. (The following list is taken from the July 1992 revision)
(Available in UK from: Microinfo Ltd., PO Box 3, Omega Park, Alton, Hants, GU34 2PG. Tel: 01420 86848; Fax: 01420 89889)

Method 1 – Sample and velocity traverses for stationary sources.
Method 1A – Sample and velocity traverses for stationary sources with small stacks or ducts.
Method 2 – Determination of stack gas velocity and volumetric flow rate (Type S pitot tube).
Method 2A – Direct measurement of gas volume through pipes and small ducts.
Method 2B – Determination of exhaust gas volume flow rate from gasoline vapor incinerators.
Method 2C – Determination of stack gas velocity and volumetric flow rate in small stacks or ducts (standard pitot tube).
Method 2D – Measurement of gas volumetric flow rates in small pipes and ducts.
Method 3 – Gas analysis for carbon dioxide oxygen, excess air, and dry molecular weight.
Method 3A – Determination of Oxygen and Carbon Dioxide Concentrations in Emissions From Stationary Sources (Instrumental Analyzer Procedure).
Method 4 – Determination of moisture content in stack gases.
Method 5 – Determination of particulate emissions from stationary sources.
Method 5A – Determination of particulate emissions from the asphalt processing and asphalt roofing industry.
Method 5B – Determination of non-sulfuric acid particulate matter from stationary sources.
Method 5D – Determination of particulate emissions from positive pressure fabric filters.
Method 5E – Determination of particulate emissions from the wool fiber-glass insulation manufacturing industry.
Method 5F – Determination of non-sulfate particulate matter from stationary sources.
Method 5G – Determination of particulate emissions from wood heaters from a dilution tunnel sampling location.
Method 5H – Determination of particulate emissions from wood heaters from a stack location.
Method 6 – Determination of sulfur dioxide emissions from stationary sources.

Method 6A – Determination of sulfur dioxide, moisture, and carbon dioxide emissions from fossil fuel combustion sources.
Method 6B – Determination of sulfur dioxide and carbon dioxide daily average emissions from fossil fuel combustion sources.
Method 6C – Determination of Sulfur Dioxide Emissions From Stationary Sources (Instrumental Analyzer Procedure).
Method 7 – Determination of nitrogen oxide emissions from stationary sources.
Method 7A – Determination of nitrogen oxide emissions from stationary sources – Ion chromatographic method.
Method 7B – Determination of nitrogen oxide emissions from stationary sources (Ultraviolet spectrophotometry).
Method 7C – Determination of nitrogen oxide emissions from stationary sources – Alkaline-permanganate/colorimetric method.
Method 7D – Determination of nitrogen oxide emissions from stationary sources – Alkaline-permanganate/ion chromatographic method.
Method 7E – Determination of Nitrogen Oxide Emissions From Stationary Sources (Instrumental Analyzer Procedure).
Method 8 – Determination of sulfuric acid mist and sulfur dioxide emissions from stationary sources.
Method 9 – Visual determination of the opacity of emissions from stationary sources.
Alternate method 1 – Determination of the opacity of emissions from stationary sources remotely by lidar.
Method 10 – Determination of carbon monoxide emissions from stationary sources.
Method 10A – Determination of carbon monoxide emissions in certifying continuous emission monitoring systems at petroleum refineries.
Method 10B – Determination of carbon monoxide emissions from stationary sources.
Method 11 – Determination of hydrogen sulfide content of fuel gas streams in petroleum refineries.
Method 12 – Determination of inorganic lead emissions from stationary sources.
Method 13A – Determination of total fluoride emissions from stationary sources – SPADNS zirconium lake method.
Method 13B – Determination of total fluoride emissions from stationary sources – Specific ion electrode method.
Method 14 – Determination of fluoride emissions from potroom roof monitors for primary aluminium plants.
Method 15 – Determination of hydrogen sulfide, carbonyl sulfide, and carbon disulfide emissions from stationary sources.
Method 15A – Determination of total reduced sulfur emissions from sulfur recovery plants in petroleum refineries.
Method 16 – Semicontinuous determination of sulfur emissions from stationary sources.
Method 16A – Determination of total reduced sulfur emissions from stationary sources (impinger technique).
Method 16B – Determination of total reduced sulfur emissions from stationary sources.
Method 17 – Determination of particulate emissions from stationary sources (instack filtration method).

Method 18 – Measurement of gaseous organic compound emissions by gas chromatography.
Method 19 – Determination of sulfur dioxide removal efficiency and particulate, sulfur dioxide and nitrogen oxides emission rates.
Method 20 – Determination of nitrogen oxides, sulfur dioxide, and diluent emissions from stationary gas turbines.
Method 21 – Determination of volatile organic compound leaks.
Method 22 – Visual determination of fugitive emissions from material sources and smoke emissions from flares.
Method 23 – Determination of Polychlorinated Dibenzo-p-Dioxins and Polychlorinated Dibenzofurans From Stationary Sources.
Method 24 – Determination of volatile matter content, water content, density, volume solids, and weight solids of surface coatings.
Method 24A – Determination of volatile matter content and density of printing inks and related coatings.
Method 25 – Determination of total gaseous non-methane organic emissions as carbon.
Method 25A – Determination of total gaseous organic concentration using a flame ionization analyzer.
Method 25B – Determination of total gaseous organic concentration using a non-dispersive infrared analyzer.
Method 26 – Determination of Hydrogen Chloride Emissions From Stationary Sources.
Method 27 – Determination of vapor tightness of gasoline delivery tank using pressure–vacuum test.
Method 28 - Certification and auditing of wood heaters.
Method 28A – Measurement of air to fuel ratio and minimum achievable burn rates for wood-fired appliances.

Hazardous Air Pollutants (CFR40, Part 61, Appendix B)
Method 101 – Determination of particulate and gaseous mercury emissions from Chlor-alkali plants – air streams.
Method 101A – Determination of gaseous and particulate mercury emissions from sewage sludge incinerators.
Method 102 – Determination of gaseous and particulate mercury from chlor-alkali plants – hydrogen streams.
Method 103 – Beryllium screening method.
Method 104 – Determination of beryllium emissions from stationary sources.
Method 106 – Determination of vinyl chloride emissions from stationary sources.
Method 108 – Determination of gaseous and particulate arsenic emissions.
Method 110 – Determination of benzene from stationary sources.
Method 111 – Determination of polonium-210 emissions from stationary sources.
Method 114 – Test method for determining radionuclide emissions from stationary sources.
Method 115 – Monitoring for radon-222 emissions.

Performance Specifications (CFR40, Part 60, Appendix A)

1. Specifications and test procedures for opacity continuous emission monitoring systems in stationary sources.

2. Specifications and test procedures for SO_2 and NO_x continuous emissions monitoring systems in stationary sources.
3. Specifications and test procedures for O_2 and CO_2 continuous emissions monitoring systems in stationary sources.
4. Specifications and test procedures for carbon monoxide continuous emission monitoring systems in stationary sources.

4A. Specifications and test procedures for carbon monoxide continuous emission monitoring systems in stationary sources.

5. Specifications and test procedures for total reduced sulphur continuous emission monitoring systems in stationary sources.
6. Specifications and test procedures for continuous emission rate monitoring systems in stationary sources.
7. Specifications and test procedures for hydrogen sulfide continuous emission monitoring systems in stationary sources.

Quality Assurance Procedures

Procedure 1. Quality assurance requirements for gas continuous emission monitoring systems used for compliance determination.

Appendix D

Main provisions for testing to British Standard 3405:1983

If the main provisions for testing to BS 3405:1983, listed below, are not fully complied with, then the accuracy of the results, from such a sampling exercise, are questionable (i.e. worse than the 25% accuracy claimed for the standard).

Sample plane

- The sampling plane is not less than one flue diameter (or hydraulic equivalent) from the nearest bend or obstruction, always providing that the pitot–static tube measurements satisfy the criteria detailed below. (A distance of four diameters minimum is preferable.)
- Duct dimensions at the sample plane are measured within 10 mm/metre.
- All sample points to be in the same sample plane, across the duct.

Pitot–static tube measurements

- No pitot–static tube pressure readings, measured at the sample plane, are zero or negative.
- The sum of the square roots of the pitot–static tube pressure readings, measured at the sample points, does not differ by more than 5% before and after isokinetic sampling.
- The direction of gas flow at any sample point shall be not more than 2° from the axis of the flue. (The ' angle of swirl ' shall not be more than 20°)
- The pitot–static tube head must not be obstructed. (i.e. a pitot–static tube heater or other device must not be used).
- If the ratio of the highest to lowest pitot–static tube readings, recorded during the preliminary 2 × 10-point pitot–static tube pressure measurement traverse, differs by more than 4 : 1, then four-point sampling will be replaced by eight-point sampling.
- The ratio of the highest to lowest pitot–static tube readings, recorded during the preliminary 2 × 10-point pitot–static tube pressure measurement traverse, does not differ by more than 9 : 1 or more. (Where the ratio of 9 : 1 is equalled or exceeded then the sample plane is unsuitable for isokinetic sampling).
- A standard NPL or calibrated pitot–static tube should normally be used to obtain all velocity pressure measurements. (However, in ducts where the gas flow contains free moisture, or the duct gases are at or near the dew-point, the use of a calibrated S-type pitot–static tube may be necessary, to overcome the problems of measurement associated with the presence of such moisture. Under these circumstances the declared results must be qualified.)

Duct and duct gas conditions

- Temperatures measured at adjacent traverse points do not differ by more than 10%.
- Duct gas temperatures can be measured within 5K.
- All temperatures measured at the temperature traverse sample points are within 10% of the mean value.
- Velocity of the duct gases at the sample plane is not below 3 m/s average, at 200°C, or is such that the pitot–static tube velocity pressure is not less than 5 Pa.
- Static pressure in the duct not to be less than 0.5 kPa below atmospheric (50 mbar vacuum).
- Duct gas must be free from water droplets, such as to permit pitot–static tube readings to be obtained. (see also comments under Pitot Static Tube measurements above.)

Sampling conditions

- Sampling is carried out at the centroid of equal areas (i.e. four-point or eight-point , measured within 20 mm/m of duct dimension.)
- Sampling is carried out isokinetically.
- The duration of sampling at any sampling point shall be not less than 3 minutes.
- Where cumulative sampling is undertaken, the product 'a × t' must be kept constant for each of the sample points. ['a' is the cross-sectional area of the sampling nozzle used and 't' is the duration of sampling, at that sample point.]
- If the duct cross-sectional area is greater than 2.5 m^2, then eight-point sampling shall replace four-point sampling.
- The temperature and pressure of all volume flow settings/measurements, required as part of the operational procedure for the sampling equipment, are so maintained throughout the sampling period, as to ensure that the actual isokinetic sample extraction rate is consistent, with the gaseous flow conditions within the duct, chimney or flue being sampled.
- Where the sampling is carried out at a single average velocity point, or at less than the minimum number of sample points required, for whatever reason, the results must be clearly qualified.
- Sampling equipment is tested leak-free, before any sampling is undertaken.
- The sampling equipment has not been subject to unqualified modifications and is operated in the approved manner, as described, in the literature associated with that equipment.

Collected particulates

- Weighing can be performed satisfactorily.
- The effect of the presence of interference materials, organics, hydrated minerals, tars, etc., in the collected particulates, are fully compensated, post sampling handling, during the conditioning and weighing procedures, so as not to materially alter the result.
- The balance used is capable of weighing to ± 0.1 mg.
- Weights of particulate sample collected, in the primary sample collector, to be not less than 0.3% of that particulate collector weight (this applies to the primary collector only). A lower particulate sample weight may be acceptable, under the circumstances of :

(a) short-duration process emission tests, as long as the isokinetic sampling period has been of at least 10 minutes duration at each point
(b) where three complete isokinetic sampling runs are undertaken, which are within the 1½ times factor* and all other provisions listed are met. (*1½ times factor. The ratio of the results for the mass emission, calculated for each separate isokinetic sampling run, shall not be more than 1.5 : 1 [Section 3. Clause 11])

- The dust collector must be capable of collecting grit and dust with an efficiency of retention, for the stated Stokes diameters of :
  (a) not less than 98% of particles larger than 20 μm;
  (b) not less than 96% of particles in the range 10 – 20 μm;
  (c) not less than 90% of particles in the range 5 – 10 μm;
  (d) not less than 60% of particles in the range 1 – 5 μm.

NB. This is satisfied by the BCURA cyclone, where grit and dust are concerned (i.e. for particles down to Stokes diameters of 1 μm)

- The sample collector for total particulates must be capable of collecting 98% of all solid particles. (NB. The back-up filter must be used with the BCURA cyclone where the particulates in the range 1 – 5μm are known to exceed 5%.)

Results

- The concentration of particulates in the emission shall be calculated from the measurements made during the sampling exercise (and the subsequent weights obtained for the particulate collection). In particular, using the values obtained for the pitot–static tube velocity pressures, measured at each of the sample points, immediately before each isokinetic sampling traverse.
- The gaseous volumes, used in the calculation of the results, will normally be corrected to the reference conditions, laid down in the appropriate Chief Inspector's Guidance Notes, for the process being tested.

Index

Page numbers appearing in **bold** refer to figures and page numbers appearing in *italics* refer to tables.